Machine Nature

Machine Nature

The Coming Age of Bio-Inspired Computing

Moshe Sipper

McGraw-Hill

New York   Chicago   San Francisco
Lisbon   London   Madrid   Mexico City   Milan
New Delhi   San Juan   Seoul   Singapore
Sydney   Toronto

Library of Congress Cataloging-in-Publication Data

Sipper, Moshe.
    Machine nature : the coming age of bio-inspired computing / Moshe
    Sipper.
        p.   cm.
    Includes bibliographical references and index.
    ISBN 0-07-138704-8 (acid-free paper)
    1. Bio-inspired computers.   I. Title.

    QA76.887 .S57 2002
    621.39'1—dc21

                                                    2002004272

McGraw-Hill

A Division of The **McGraw·Hill** Companies

1 2 3 4 5 6 7 8 9 0   DOC/DOC   0 8 7 6 5 4 3 2

0-07-138704-8

 This book is printed on recycled, acid-free paper containing a minimum of
50% recycled de-inked fiber.

I never think of the future. It comes soon enough.
Albert Einstein

In memory of my cousin, Tami,
who died so unjustly soon.

Contents

Acknowledgments

It is a pure joy to have so many friends and colleagues to whom I am indebted.

Eduardo Sanchez of the Swiss Federal Institute of Technology in Lausanne was my first reader and critic, whose enthusiasm and encouragement were a *sine qua non* for my deciding to pursue this undertaking.

Marco Tomassini of the University of Lausanne, Switzerland, was brave enough to work his way through an entire first draft! His remarks and encouragement were invaluable to me.

Daniel Mange of the Swiss Federal Institute of Technology in Lausanne read a large portion of the first draft, and his downright positive attitude was a boon to my spirits.

To you, *mes amis*, I wish to add this: Your contribution to this book goes beyond the mere critiques of the text, which you so handsomely provided. "If I have seen further it is by standing on the shoulders of giants," wrote Sir Isaac Newton. Daniel, Eduardo, and Marco: Thank you for being such amicable giants.

My thanks to Edmund Ronald: Sometimes trying, often challenging, always a friend—and a true *aide-monde*.

I thank André ("Chico") Badertscher for all those magnificent photographs, as well as for the wonderful bouts of *jeux de mots*, which have helped me sharpen my French. *Merci, cher Baderts.*

The Logic Systems Laboratory (LSL) at the Swiss Federal Institute of Technology in Lausanne has been an ideal environment for doing research, combining both keen minds and lively spirits. I wish to thank each and every one of my colleagues at the LSL.

Acknowledgments

I wish to thank wholeheartedly the following people for their advice, encouragement, remarks, and suggestions: Forrest Bennett, Dario Floreano, Max Garzon, John Koza, Carlos-Andrés Peña-Reyes, Andrés Pérez-Uribe, Anne Renaud, Andrea Tettamanzi, and Christof Teuscher.

Last, but not least, I thank my editors at McGraw-Hill, Kip Hakala and Ruth Mannino, who led me unto the promised land of publication.

Adaptation, Bio-Inspiration, Complexity: A New Computing ABC

Many people feel they were born too late. Then there are those who deem themselves to have been born too early; I for one belong to this latter group. I like to envision a future in which our bodies are on a par with our imagination, where humans will have unchained their earthly shackles, and, perhaps most importantly, a future in which humanity's spirit finally matches its technological wizardry.

Maybe that is why my research revolves around what might, prima facie, seem like science fiction: machines and computers that adapt, evolve, learn, heal, reason, and more—accomplishing feats that we usually associate only with Nature. But this seemingly fictional science is in fact quite real.

Much of the work is carried out by daring, creative researchers who incessantly push at the frontiers of knowledge, with some of their ideas having already found their way into products such as automobiles and washing machines.

During the past few years a new wind has been sweeping through the computing terrain, slowly changing our fundamental view of computers. We want them to be faster, better, more efficient—and proficient—at their tasks. But that's just part of the story. The other part, and in my mind the more exciting, is that we've come to expect computers to stop being so *stiff*.

Computers are rigid, unbending, unyielding, inflexible, and quite unwieldy. Let's face it. They've improved our lives in many a way, but they do tend to be a pain. When interacting with them, you have to be very methodical and precise, in a manner quite contrary to human nature. Step outside the computer's programmed repertoire of behavior, and it will simply refuse to cooperate, or—even worse—it will "crash" (a vivid term coined by computing professionals to describe a computer's breaking down). Computers are notoriously bad at learning new things and at dealing with new situations. It all adds up to one thing: At their most fundamental, computers lack the ability to *adapt*.

Adaptation concerns a system's ability to undergo modifications according to changing circumstances, thus ensuring its continued functionality. We often speak of an environment and of the system's adjustment to changing environmental conditions. The archetypal examples of adaptive systems are not among human creations, but among Nature's. From bacteria to bumblebees, natural organisms show a striking capacity to adapt to changing circumstances, a quality that has not escaped the eyes of computing scientists and engineers. The influence of the biological sciences in computing is on the rise, slowly but surely inching its way toward the mainstream. There are many examples today of systems inspired by biology, known as *bio-inspired* systems.

These adaptive, bio-inspired systems are *complex*, which refers to more than their simply being complicated objects or to the difficulty of building and comprehending them. As Peter Coveney and Roger Highfield wrote in *Frontiers of Complexity*: "Within science, complexity is a watchword for a new way of thinking about the collective behavior of many basic but interacting units, be they atoms, molecules, neurons, or bits within a computer. To be more precise, our definition is that *complexity is the study of the behavior of macroscopic collections of such units that are endowed with the potential to evolve in time.* Their interactions lead to coherent collective phenomena, so-called emergent properties that can be described only at higher levels than those of the individual units. In this sense, the whole is more than the sum of its components...."

Natural organisms are complex, adaptive systems, and our artifacts are now beginning to follow in their footsteps. Adaptation, Bio-inspiration, and Complexity thus underlie the new computing ABC.

"Much have I travell'd in the realms of gold, And many goodly states and kingdoms seen," wrote John Keats. In the next chapter we begin our voyage to the frontiers of computing, a veritable journey into realms of gold. Each chapter adds a bead to our ABC chain, a chain that defines what I call the *Terra Nova* of computing, where computers do all those things we'd *like* them to do:

- Darwinian evolution occurs only in nature? No longer. In Chapter 1 we will see how engineers use evolution to create not only new objects, but indeed an entirely new *way of creating objects.*
- Are computer programmers industrious? Not all of them. As we'll see in Chapter 2, some of them are quite lazy, reclining comfortably in their swivel chairs while evolution does all the work: evolving computer programs.
- "Why doesn't a computer *ever* learn?" you've probably asked yourself 53 times over the past week alone. As we'll see in Chapter 3, they actually *can* learn, given the right incentive.
- Why do R2D2 and C3PO exist only on the silver screen? In Chapter 4 we'll see what researchers in adaptive robotics are doing to correct this unfortunate situation.
- Why doesn't your computer understand such a simple phrase as "It's a bit warm in here"? In Chapter 5 we consider computers that can talk and reason in a fuzzy manner, able to come to grips with phrases such as "a bit" and "warm"—just like us.
- Computer hardware is too hard and unyielding? In Chapter 6 we'll meet the latest in computer-chip technology: soft hardware, namely, computer chips that are malleable.
- Can we do something about computers breaking down much more easily than we do? I shudder to

think what would happen if I were to crash three times a day. And even when we do crash, we humans have wonderful bodies that are often able to heal. As we'll learn in Chapter 7, computers are beginning to experience the joys of healing.

- Wouldn't it be nice to equip your computer with an immune system to ward off all those nasty viruses lurking in the dark corners of cyberspace? In Chapter 8 we talk about immune systems for computers.
- Will silicon circuits ever behave like carbon beings? This fundamental gap between the two (somewhat less obvious in California) is starting to close with the advent of DNA computing, a topic we encounter in Chapter 9.
- Your body is made up of gazillions of tiny cells that work in concert to produce the symphony you call "I." In Chapter 10 we listen to the music of cellular computers.

These fields of research and application are not studied in isolation from one another; just like their living counterparts, they too interact. We'll see how we can combine several of the above approaches—such as evolution, learning, and fuzziness—within a single human-made system, thus greatly enhancing its performance.

With our newly won knowledge of all these novel lands in the *Terra Nova* of computing, it is time to dig a little deeper:

- In Chapter 11 we ask: What are the fundamental differences between Nature-made and human-made systems?

- If we let evolution run loose in a computer, then who controls the process? Who's the boss? We tackle that one in Chapter 12.
- What do you get when you cross a scientist and an engineer? A scigineer, of course. We'll meet this hybrid fellow in Chapter 13.
- Can our creations one day take on a life of their own? Now there's a truly *big question*. You'll have to wait till Chapter 14 to learn more about this one.
- Finally, in Chapter 15, we pick up where we started here, going from ABC all the way up to Z.

This book aims to convey more than just such catch phrases as "computers can evolve" or "machines can heal." My aim is to provide you not only with the "what" but also with a basic inkling of the "how," not only to present you with front-page headlines ("Researchers Fabricate Novel Computers Based on DNA!") but also to describe in lay terms just *how* such computers work. With luck, this will replace the uneasy "pie-in-the-sky" feeling usually induced by such slogans, with an understanding—and thus an appreciation—of the real strides that are continually being made, as well as of the many open questions that still remain.

I hold the strong conviction that research should not be judged by its immediate applicability. True, some projects will find their way into commercial products almost immediately. Yet the seeds we sow today may give rise to marvelous trees of knowledge 20 or 30 years hence. And I believe firmly that society at large will benefit from such an outlook. I cannot help but be reminded of the story of William Gladstone who, as Chancellor of the Exchequer, was invited to a demonstration of Michael Faraday's equipment for generating the latest scientific wonder at the time—electricity. Faraday set up the experiment and ran it, while Gladstone looked coolly on. After the demonstration, Gladstone stood silent for a moment, and then said: "It is very interesting, Mr. Faraday, but what practical worth is it?" Undaunted, Faraday replied: "One day, sir, you may tax it." My research-for-tomorrow position notwithstanding, most of the topics covered in this book are studied today both by cutting-edge researchers and by stern industrialists.

While the devil may be in the details, I've tried to be angelic by writing about the important, basic principles without any unexplained technical jargon. Also, this book tries to appeal to your imagination as much as to your reason. As Albert Einstein said, "Imagination is more important than knowledge."

Before we begin our journey, a small remark about sex. One of the problems in modern English writing concerns the use of the gender pronouns. I've applied Occam's razor, opting for the simplest solution: I use one or the other entirely at whim.

In *Through the Looking-Glass* Lewis Carroll wrote:

> *"The time has come," the Walrus said,*
> *"To talk of many things:*
> *Of shoes—and ships—and sealing wax—*
> *Of cabbages—and kings—*
> *And why the sea is boiling hot—*
> *And whether pigs have wings."*

Indeed.

CHAPTER 1

Darwin's Way

It's your first day at your new job. You're quite proud of yourself—and rightly so. Those long, hard years at that top-notch engineering school, the sleepless nights before exams you spent in a last-minute attempt to digest all that material, not to mention your feeble excuse for a social life. But all that's over. You're now a proud member of the engineering clan. The first week is okay: You are shuffled around, introduced to your new coworkers, given the rundown on the company's business, in short, the standard settling-in tour.

Then comes that dreaded day when your new boss calls you to her office and hands you your first assignment. The fun's over. Now you're asked to actually do something. "Well," she says, "this is a tough one. Our firm has been commissioned to design a bridge across Lake Geneva between the Swiss town of Lausanne and the French town of Evian. Now, the distance poses no problem: It's about 10 kilometers in all, and building such bridges is a cinch. You probably designed such stuff during your freshman year. But here's the catch: These guys want a retractable bridge! They want to be able to fold and unfold it—at whim—in a matter of hours. Apparently, they'll put it up when there's an overflow of tourists, and then retract it when the party's over."

"I must admit," continues your boss, "that we were quite flabbergasted when this request was made of us; but we agreed to look into it. That was six months ago, and I can tell you straight away that we are at a complete standstill. I've put the best engineering minds in our firm on this problem, and the results are zilch. That's where you come in. Since you're fresh out of school, I've decided to let you have a go at it. I know this sounds a bit paradoxical at first, but I'm actually counting on your fresh attitude and your being unencumbered by too much past experience. And your being full of all that state-of-the-art engineering lore might just act to tip the scales in our favor.

"So," she ends, handing you a thick wad of paper, "it's up to you now. Don't feel too bad if you can't solve this. After all, we're engineers, not magicians. In any case, good luck!"

So, how can you go about designing such a demanding bridge? After all, as your boss pointed out, the best engineering minds in the firm came up empty-handed. Yet, you think to yourself, most engineering design involves modifications within an existing framework. Those other engineers probably tried to apply classical, well-known bridge-design practices, overstretching them to the limit, and to no avail. This retractable bridge is something entirely new, something that calls for a *really* novel design. It is a new *kind* of bridge, a new *species*. That's when it hits you: Charles Darwin.

In 1859 Charles Darwin published his masterpiece *On the Origin of Species by Means of Natural Selection, or the Preservation of Favoured Races in the Struggle for Life.* Though science has by now come to terms with it, the shock waves produced by Darwin's revolutionary theory of evolution are still reverberating within society at large. In one fell swoop Darwin had further removed *Homo sapiens* from their self-proclaimed lofty position in the universe: In the sixteenth century, Copernicus and Galileo displaced us from our pedestal at the center of the universe to that of a mere ball of mud circling a fiery furnace in space. We thought, then, that at least there was no doubt as to our central position here on *our* planet, high and above all those other creatures. Sorry, proclaimed Darwin three centuries later, we're actually but one branch in the grand Tree of Life. We are but another element of the vast richness of flora and fauna that have evolved over the eons. What's more, Darwin also revealed to us that those amusing, tree-swinging animals are actually quite close relatives of ours.

Natural evolution is a powerful force. It has brought about a plethora of complex machines, such as eyes, wings, and brains, the construction of which are still well beyond our current engineering capabilities. Nature is the most august engineer known to humans—though a blind one at that. Her designs come into existence through the slow process, over millions of years, known as evolution by natural selection. As beautifully enunciated by Darwin,

> If during the long course of ages and under varying conditions of life, organic beings vary at all in the several parts of their organisation, and I think this cannot be disputed; if there be, owing to the high geometrical powers of increase of each species, at some age, season, or year, a severe struggle for life, and this certainly cannot be disputed; then, considering the infinite complexity of the relations of all organic beings to each other and to their conditions

of existence, causing an infinite diversity in structure, constitution, and habits, to be advantageous to them, I think it would be a most extraordinary fact if no variation ever had occurred useful to each being's own welfare, in the same way as so many variations have occurred useful to man. But if variations useful to any organic being do occur, assuredly individuals thus characterised will have the best chance of being preserved in the struggle for life; and from the strong principle of inheritance they will tend to produce offspring similarly characterised. This principle of preservation, I have called, for the sake of brevity, Natural Selection.

The evolutionary process can be seen to rest on four principles:

- Individual organisms vary in their viability in the environments that they occupy.
- This variation is heritable.
- Individuals tend to produce more offspring than can survive on the limited resources available in the environment.
- Those individuals best adapted to the environment will survive to reproduce in the ensuing struggle for survival.

The continual workings of this process over the millennia cause populations of organisms to change and generally become better adapted to their environment.

This is, of course, but a high-level abstraction that hides all those fascinating biological mechanisms that have remained under intense investigation to this day. We still have much to learn in the field of biological evolution. Abstraction, though, is the bread and butter of computing scientists; they like to extract principles, ignoring the nitty-gritty, thereby coming up with entirely new ideas for computation, for example, evolutionary computation.

Starting in the 1950s, about a century after the publication of Darwin's *Origin of Species*, several researchers from Australia, England, Germany, and the United States came up with an ingenious idea: implementing within a computer a simulacrum of evolution by natural selection. They reasoned that if natural evolution can produce such complex organisms, perhaps artificial evolution could be put to use in producing engineering designs for problems that are beyond the reach of classical methodologies.

So how does one go about simulating evolution within a computer? All we have to do is program a virtual environment governed by the four principles stated above. There is, however, one crucial difference between natural evolution and artificial evolution that must be borne in mind: The former is an open-ended process, while the latter is a guided one. This means that the evolutionary scenario in nature is highly complex, and we would be hard put to define in precise terms what exactly enables one organism to survive and produce a multitude of offspring, while another dies childless. That is to say, the survival criteria in nature are quite elaborate, involving an intricate interplay between the organism's external environment—temperature, humidity, altitude, and so on—and its cohorts, other living beings with which it shares an ecological niche (be they friend or foe). On the other hand, the paradigm known as evolutionary computation is at heart an engineering methodology, meaning that we are trying to *solve a given problem*; thus, the engineer imposes selection criteria in accordance with the problem at hand. While natural evolution, as seen by science, admits no "hand of god," with the survival criteria emerging from the multitude of interacting organisms, simulated evolution does admit such a hand, with "god" being the engineer who guides the process of finding a solution to a given problem.

Let's go back now to the aspiring young engineer in search of a solution to the retractable bridge problem. Luckily, you took that elective course on evolutionary computation in your senior year. "Great," you think to yourself, "maybe Darwin's way holds the solution to my problem." The first thing you need to do is to create a *population* of *organisms*. There are no biological details here, just an *analogy* with the biological process of evolution. In your case the "species" in question is definitely of the artificial kind: It comprises bridges, actually a population of "bridge organisms."

Let's take one more step into biological territory and talk about genotypes and phenotypes. This is an important distinction in nature: An organism's genotype is its genetic constitution, the chain of deoxyribonucleic acid (DNA) that contains the instructions necessary for the making of the individual. In humans, this DNA sequence comprises some three-billion-odd letters. The genotype (often referred to as the genome) is neatly tucked away, safe and sound in the insides of each and every cell of your body. But it is not the genotype that "struts and frets his hour upon the stage," to use Shakespeare's immortal words from *Macbeth*. That is the role of the phenotype, the mature organism that emerges through execution of the instructions written in the genotype. It is the phenotype that is subjected to the fierce battle for survival. Your genotype is safe and snug in your cells, while you—the phenotype—are the one to face the elements. But the last word belongs to the genotype. It alone carries the accumulated evolutionary benefits over to future generations.

What—if anything—has this to do with our bridge organisms? Well, they too exhibit this genotype-phenotype duality, and it is up to you as an engineer to set it up. The genotype of a bridge organism contains the instructions on how to build a bridge. The phenotype is the bridge built by following the genotypic instructions. Designing the exact makeup of the genotype occupies you for the next few weeks—not bad considering that Nature took a billion-odd years to design her first genotype.

What does the genome (genotype) of the bridge organism look like? First of all, it is *generic*, meaning that it does not define one single individual but rather a family (or species) of individuals. *Homo sapiens* are a good example: We all possess the same basic genotypic structure, with individuals differing as a consequence of small differences within this framework. For example, you might decide that a bridge will consist of ten bricks, with its genotype specified by ten spatial coordinates—one per brick. This is a generic genotype, because by plugging in different ten-number sequences, you end up with different bridges. Obviously, your ultimate genotype will be much more complicated than this since you can't get very far with ten bricks. In all likelihood the final genome will not be a mere list of spatial coordinates for brick laying; in fact, the more generic your bridge organism, the larger the range of possible bridge individuals. This means that you will have more of a chance of finding a good bridge, but there will also be many more bad ones out there, and in fact many of these will turn out to be downright horrendous. The latter is related to what is known in evolutionary computation parlance as the *search space*.

What is a search space? Consider again the ten-brick bridge organism specified by a genotype of ten spatial coordinates. Suppose that each spatial coordinate can take on 1 of 100 possible values (for example, only specific coordinates within the designated Lake Geneva area). How many *possible* genotypes are there? To obtain the answer, you have to multiply the number of possible values for the first brick (100) times the number of possible values for the second brick (100), and so on until the tenth brick. In mathematical terms this amounts to 100 raised to the power of 10, or 100,000,000,000,000,000,000. In so-called search problems, where one is basically searching for a good solution among many possibilities, it is often customary to invoke an imaginary "space" in which each possible solution (or genotype in our case) is represented by a point. The size of the search space is then simply the number of such points.

Search spaces for real-world problems are usually huge. The above ten-brick genotype, with 100 possible values per brick coordinate, is actually quite small as search spaces go. Despite its relative smallness, even if you could examine, say, 100 genotypes per second, going through the whole search space would still take you about 30 billion years (roughly twice the age of the universe). Search-space sizes for the vast majority of real-world problems are notorious for producing such "ridiculous" numbers.

The bad news is that you cannot examine every possible bridge genotype. The good news is that you don't have to; that's where evolution steps in. Whether natural or artificial, the evolutionary process does not work by trying out each and every possible genotype. In your simulacrum of evolution, you set out with a population of, say, 500 individual organisms (genotypes) that are created *completely at random* (even the lowliest personal computer can easily churn out such random individuals). Now comes a crucial point: Each genotype gives rise to a unique phenotype that must then engage in the battle for survival; it must prove its *fitness.* The environment in your simulation is quite simple: It consists solely of bridges. It remains for *you*—the "hand of god"—to decide what defines a "fit" bridge. Being the talented engineer that you are, you manage to come up with a good fitness definition. This is far from easy. You have to consider all the criteria that the intended bridge must satisfy: It must be stable; it must be robust (able to handle extreme weather conditions); it should be small; preferably it will not be too expensive (obviously, ten bricks will always cost the same, but of course your ultimate genotype is much more complex than that!); and it also must not only be retractable, but the retraction should take as short a time as possible. All these criteria, among others, go into your fitness definition. This is one of your hardest tasks in applying the evolutionary-computation methodology; not only do you have to identify all the relevant fitness criteria, but you must also define their relative importance (for example, stability is probably more important than cost).

The wheels have now been set in motion for commencing this simulated evolution. From here on it's hands-off for you. The computer will now automatically run the evolutionary process. It begins with the population of 500 random bridge genotypes—each representing a possible solution to the bridge problem—collectively referred to as the first generation. Each of these genotypes is now translated into a phenotype. This can be done, say, by building a real-world model or, more likely, by running a computer simulation of the workings of the bridge in question. Each phenotype now takes the "survival" test, which simply means that it receives its own personal score, representing to what degree it fulfills the fitness definition. A higher score implies better fitness, which in turn means that the underlying genotype is better.

The individuals of this first generation will usually turn out to be very bad, that is, they will exhibit very low fitness. What did you expect? After all, the computer simply picked 500 random genotypes out of a huge search space. Some of these organisms may be entirely unfit, representing bridges that cannot be built at all! Imagine a ten-brick genotype with two overlapping coordinates: Obviously, two bricks cannot occupy the same physical space. Thus, such a solution is illegal, or it has a fitness of zero. What you will find in this first random generation is *variety*: Different organisms will earn different fitness scores. Slight as they may be, even minute differences are sufficient to drive the evolutionary process.

The computer now picks out those individuals that will be allowed to reproduce, to leave offspring in the next generation. This is done by holding a sort of lottery, where better-adapted individuals—ones with higher fitness—get a larger share of the available lottery tickets, only a fixed number of which are available. Thus, the "stronger" will tend to survive, though the "weak" are not necessarily chucked out automatically; they simply have a lower chance of winning the who-gets-to-reproduce lottery. (We get back to why keep the weak ones at all further on.) Now you're left with more of the good individuals and fewer of the bad ones, though you still have nothing *new*. All you have is a different mixture of those first-generation organisms. It's time for sex.

Having selected who gets to reproduce and how often, pairs of lucky moms and dads now come together and mate. They exchange parts of their genetic material, each parent donating a piece of its genome to the other. Borrowing from molecular biology, sex in evolutionary computation is mundanely called *crossover*. And, as if this merry mating festival were not enough, the offspring now get "mutated" here and there: Small random changes are made to their genomes. Again, this idea is borrowed from biology. From cosmic rays to copying errors, the reproductive process is not perfect, and mutations arise from time to time (though usually at a low rate).

Selection of the fittest by a biased lottery, followed by snappy marriages and zealous matings, ending with a sprinkle of mutations: Behold the new generation of bridge organisms. Now comes the easy part. The computer simply keeps iterating the process: evaluating the new generation, that is, assigning a fitness score to each new individual, followed by a biased lottery, where fitter organisms hold more "tickets," after which the winners zip off to their weddings, where crossovers and mutations are the rule.

All in all it seems rather straightforward so far. But does it work? Surprisingly it does. Or is it such a surprise? After all, this simple simulacrum of evolution does have all the necessary ingredients: the four principles listed above; variability between organisms, which is heritable; and a struggle for survival (since only a limited number can make it to the next generation) in which the better adapted (with higher fitness) tend to survive.

In nature, the weak stand a chance and may sometimes survive. In the simulated scenario, you could have simply ousted them right off. After all, it's quite easy to get rid of those organisms with the lowest fitness scores. Then why hold a lottery (even if it is a biased one)? Why give the weak a chance at all? Even a weakling might possess some important trait in its genome, which may, several generations down the line, combine with other organisms' genetic material to produce a superfit individual. While granddad may be a weakling, the grandson might turn out to be as strong as a lion. For example, a very expensive bridge (by dint of which its fitness is low) may contain some nice "trick" in its genome that results in a lighter bridge. This trait might find its way through the generations into a genome of a less expensive bridge, thereby resulting in a design that is both cheaper and lighter, exhibiting higher fitness than both its ancestors.

The combination of traits is at the heart of both the natural and simulated evolutionary process. The artificial genome is usually composed of many smaller elements, referred to as *genes*. In molecular biology a gene is a short chain of nucleotides in the DNA that specifies the structure of a particular protein. The artificial gene is a small piece of the genome that is considered the elemental unit of inheritance. It is these genes that are exchanged during the mating (crossover) festival. A gene may take on one of several values (known in biology as alleles). In the ten-brick genome, for example, there are ten genes, each of which can take on one of 100 possible values (the alleles).

Evolution derives its force from the ability to combine genes. The phenotypes compete, with the winners allowed to exchange genes among themselves. The crucial point is that even mediocre individuals may have good offspring. In a yin-yang fashion, even bad guys may have some good in them. Even low-fitness organisms may possess some good genes that ultimately get combined with other good genes, creating a high-fitness individual.

Your computer is now working relentlessly, carrying out this evolution-in-a-bottle, and after three weeks of nerve-wracking waiting, the big day finally arrives: Stepping into the office you find the computer screen cheerfully bleeping, announcing to the whole world that a perfectly good solution has been found. The retractable bridge can be built after all—a (r)evolution in bridge technology.

The next time you cross a bridge, you might well ask yourself how it was designed. For the moment, the response would probably be "by using a nonevolutionary approach." But if the bridge in question is a model bridge, then Darwin does have a say.

Pablo Funes and Jordan Pollack from Brandeis University in Massachusetts used simulated Darwinian evolution to evolve Lego™ bridges. They began by incorporating a model of the physical properties of Lego-built structures within their simulated environment. When you join two bricks together, the force of their union depends on their shapes and on the number of knobs that join them. Funes and Pollack measured the resistance to stress for all types of brick pairs available in their Lego kit. They then wrote a program that could compute whether a bridge would hold or break, just by examining the structure's plan. The computer could thus evaluate a virtual structure without the need to actually build it, saving both time and money.

The Brandeis researchers were now ready to set up their evolutionary scenario. First, they defined the bridge genome, which is actually the specification of the network of bricks. Next, they decided how the genetic operators of crossover and mutation would be affected. Finally, they defined how to evaluate the fitness of a bridge according to two criteria: (1) Does the bridge hold? (2) With its base affixed to a Lego plate, can the bridge reach a given target point in space? The nearer the bridge is to the target point, the higher its fitness (provided, of course, it holds). Having defined the genome, the genetic operators (crossover and mutation), and the fitness, they let evolution run its course—within the computer.

What emerged out of this simulated evolutionary process were successful bridge *plans*, schematic diagrams of bridge assemblies. Funes and Pollack now faced the moment of truth: Having an evolved design in hand, would the bridge hold once they actually constructed it out of real Lego bricks? Lo and behold, it held! Figure 1-1 is an example of a "long bridge," where the target point was set far away from the base.

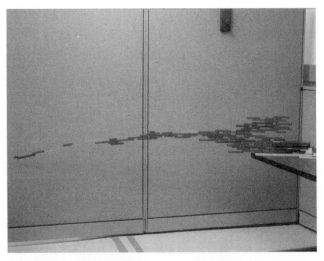

Figure 1-1   Evolved bridge
© 1998 PABLO FUNES AND JORDAN POLLACK; REPRINTED WITH THEIR KIND PERMISSION.

This bridge has somewhat of an eerie look about it, like something nature or perhaps an artist would design. We often think of art and engineering as representing two opposing ends of human creativity: To use a coarse brush, art is about creating beauty, while engineering is about creating utility. But they do have at least one thing in common: Both can benefit from Darwinian evolution. Now that we've seen evolutionary engineering at work, it is time to talk about evolutionary art.

In 1991, Karl Sims, who now heads GenArts, Inc., published an article entitled "Artificial Evolution for Computer Graphics," launching the field of evolutionary art by showing how evolution could be used to create striking digital pictures. A computer image is composed of many small dots, known as picture elements or pixels. A modern-day two-dimensional computer screen contains about one million pixels. Each such pixel can take on one of several colors (from 256 colors in low-end screens up to sixteen million colors in high-end ones). These two parameters—total number of pixels and number of possible colors per pixel—determine what is known as the screen's resolution.

To obtain a computer image, all you need to do is specify the color each pixel assumes. With your favorite graphic-design tool, which you use to draw diagrams and figures, you usually don't go all the way down to the pixel level. Popular tools allow you to draw lines, rectangles, and more complex forms without having to "paint" every pixel separately. But most such tools *do* allow you to access every single pixel if you need to fine-tune the image.

To go about evolving computer images, first we must define the picture genome. A straightforward way of doing this is to specify the color of every single image pixel. This would result, however, in a huge genome and would not allow us to advance beyond the level of a two-year-old doodler. Sims defined a more sophisticated genome, which was also much more compact: an equation that calculates the value of each pixel in accordance with its coordinates. Every pixel of our two-dimensional image has a longitude and a latitude coordinate. Think of this as a giant chessboard where each square is a pixel: The bottom-leftmost pixel has coordinates {0,0}, the one to its immediate right has coordinates {0,1}, the one immediately on top has coordinates {1,0}, and so on.

You can now use these coordinates in an equation that calculates the color of each pixel. Here's a simple equation: Calculate the sum of both coordinates, with the result taken as the pixel's color. The bottom leftmost pixel will thus have color '0,' and its two neighbors—{0,1} and {1,0}—will both have the same color, '1.' In this manner we can easily compute a color value for every single pixel of the image by simply summing the two coordinates. The computer can now display this image since it knows the colors of all the pixels. (There is still the matter of translating the color number into an actual color, but this is a trifling technicality.)

Summing both coordinate values is a simple equation that results in a digital image, probably a very dull one. But we now have a way of defining images using such compact equations without the need to specify explicitly each and every pixel: We have our picture genome. You can now create more complex equations. For example, multiply the first coordinate by two, multiply the second coordinate by seven, and then add the result of both multiplications. This is a more complicated equation, which will result in an entirely different image since each pixel will get assigned a color that differs greatly from that simple sum-both-coordinates equation.

We have no intention, however, of hand-tailoring these image equations; they are, after all, our picture genomes, so we will let them evolve. Sims did this by first generating a population of random equations, and hence of random images, which were displayed on the screen. This was the first generation of individuals in the evolutionary-art setup. To drive evolution, we now need a fitness function, in this case a measure of a picture's beauty. How does a computer know beautiful from ugly? It doesn't. Enter the "hand of god": you. Assigning a numeric score to each picture, you determine the images' fitness values. The computer now selects the best of the lot, those images with the highest fitness scores, and generates automatically the next generation. It does this by crossing over and mutating the underlying genomes. Once this new generation of genomes is available, the computer displays the corresponding phenotypes, the images themselves. This process now goes on: You assign to each image a fitness score, and the computer creates a new generation by crossing over and mutating your favorite pictures. Note that while you work at the phenotypic level, the computer handles the messy genotypic equations. All you have to do is concentrate on beauty.

The image equations get more and more complex with each generation. But does beauty emerge? That depends on the beholder of course, though I personally beheld magnificent images, one of which is reprinted in the Coda.

I cannot help but wonder what Darwin would have thought upon seeing himself turned into a bridge between engineering and art.

CHAPTER 2

Lazy Programming

I t's the night of the big match. You and a select group of friends gather around the television set, solemnly bowing before the gods of football in a ceremony reminiscent of the cave days of yore, when our ancestors would cluster around the blazing fire. The fridge is full to the brim. Your kitchen has been transformed into a railway depot; trains loaded with goods constantly depart toward Grand Central Station—the living room. You remove those sumptuous-looking cheese hors d'oeuvres from the refrigerator—the ones you bought earlier at that new French deli around the corner—and place them in the brand-new microwave oven. Then it hits you: You have absolutely no idea how to use the infernal machine! It's one of the most sophisticated models the store had in stock, but since buying it yesterday, you haven't found the time to read the manual. The howls of rage from the living room—the famished tribe you have now failed—send shivers through your body. There is a happy ending, though: You still haven't thrown away your old oven.

Almost all our modern appliances, from television sets to coffeemakers, come with thick instruction manuals. The machine is a mere chunk of metal and plastic—a piece of hardware—which will sit proudly in its allotted spot in your house and do absolutely nothing at all; nothing, that is, until you've chewed your way through that manual. The appliance, that is, the hardware, needs to be *programmed,* and the manual transforms you into a *programmer* for the particular appliance in question. Who hasn't gone through this painful video experience, virtually a rite of passage these days: You spend hours learning how to program the expensive new videotape recorder, after which you use your newly gained skills to instruct the machine to record that precious episode —the only one missing in your collection, aired tonight for the first and last time. Arriving late at night after that big family dinner, you rush to the TV set, rewind the tape, cheerfully recline in your favorite sofa—only to watch the recorded 11 o'clock news....

A modern home appliance is in fact a small computer (or, to be more precise, it contains one). Just as with its big brother—the "real" computer—there are two separate levels on which it exists: the hardware, that is, the machine itself, and the software, that is, its programming. The hardware is but an inert chunk of material poised to do your bidding. It has been designed by engineers to exhibit a wide range of functionalities, but you—the programmer—must make your specific wishes known. You must supply the machine with instructions—a program—as to the precise task it is to carry out (for example, record a specific channel, at a given hour, for a specific amount of time). Where home appliances are concerned, becoming a programmer is usually a matter of a few hours of reading the manual and fiddling with all the switches. Where computers in general are concerned, becoming a computer programmer usually takes at least a few months, and perhaps even three to four years if you want to get a degree in computing science.

The vast majority of computing professionals today are concerned with the software side of the business. Only a minute percentage of the industry deals with the internal design of the computer's innards: the hardware. These hardware engineers imbue their machines with the ability to act as *general-purpose* computers, meaning that they carry out not one single task, but rather a whole variety of tasks—provided the correct program is fed in. Once the new computer is out there, its hardware changes no more (unless it's a configurable processor, which we'll encounter in Chapter 6), and it is now that the programmers' job begins. They are the ones to write all those wonderful programs that enable the computer to run the gamut from games to networking, traversing along the way word processing, account balancing, and what have you. Though the bulk of the software is written by professional programmers, you'd be surprised to learn that virtually anyone who uses a computer is somewhat of a programmer. For example, using a word processor usually involves setting various parameters such as font size, line spacing, and so on; by doing so, the user is in fact placing the finishing touches on the word processing software, which is not shipped out in a hermetically sealed state, but rather accommodates such small touches—or programming—by the user.

Programming a computer is as much an art as it is a science. Indeed, it is rather like painting in a way. Anyone can take a course and learn the basics; many will then be able to do a half-decent job, some will actually be quite good, and a tiny minority will turn out to be sensational—the new

Van Goghs. The similarity between painting and programming lies in both being based on acquirable skills as well as on ethereal ones—such as creativity, talent, and drive. The programming profession has its own Van Goghs.

So why is programming hard? As noted earlier, a computer is a general-purpose machine, capable of doing many things—provided we tell it exactly how to perform the desired task, using the computer's own language. Human beings communicate by employing prodigiously rich languages, incorporating ambiguity, large vocabularies, and complex linguistic structures (indeed, language is often cited as the prime characteristic that sets *Homo sapiens* apart from all other earthly species). Computers, on the other hand, exhibit rather poor linguistic skills, able to cope with but restricted vocabularies, simple linguistic structures, and no ambiguity whatsoever. Ideally, when a human programmer wishes to instruct the computer as to its mission (that is, to program it), he should be able to do so using a human language. However, this is not possible to date, and therefore we resort to using the computer's language; this is known as writing a program.

A program is a sequence of instructions the execution of which accomplishes the desired task (for example, right now I'm using a word processing program to write these lines). At the lowest level, these instructions are long sequences in a number sequence containing only the numbers 0 and 1. In the early days of computing, programmers had to program at this level, composing their symphonies using but these two notes. Computing professionals realized rather quickly, though, that this is too tedious a job, and they came up with the idea of *high-level languages.* While still very restrictive (unlike human tongues), such high-level languages are, nonetheless, much easier to use. The only catch is that a program written in a high-level language must be translated into the 0-1 tongue; however, this is done automatically nowadays by the computer itself. (This translation is carried out by a special program, known as a compiler, which is analogous to the aide-de-camp of an army general. The general snaps a concise, high-level command, say, "Move division 2 to zone C," and the aide-de-camp then compiles this into a series of low-level orders to the various army units involved. For example, the engineering unit will receive orders to prepare the terrain, and the transport unit will receive orders to outfit the requisite vehicles.)

Every high-level language, such as FORTRAN, BASIC, and Java, includes a small number of instructions, also known as commands. In addition, the language embodies rules on how to assemble these commands in order to obtain a legal program (where "legal" means a program that will be understood and executed by the computer). The repertoire of instructions usually includes mathematical operations (such as add, subtract, multiply, and divide), textual operations (copy text, search for words in text, and so on), user-interface commands (read what has just been typed in on the keyboard, print a message onto the screen), and control instructions. The last are worthy of special consideration since they are the "culprits" responsible for the complexity of programming (and programs). Control commands do not interact with the person sitting in front of the screen, nor do they carry out any specific computation (such as adding two numbers); rather, they form the program's "management board," directing the internal flow of information.

To see what this means, let's consider two ubiquitous control commands that are extant in virtually all current-day programming languages: *conditional branch* and *repeat*. The former command is basically a question that is posed so as to decide which part of the program to execute next. Consider a program that computes the amount of income tax to be paid by the taxpayer, and let's assume for simplicity that there are three tax categories: below $2000 per month, no income tax paid, 0 percent; between $2000 and $6000, 20 percent; and above $6000, 30 percent. The program will contain a conditional-branch command that steers the computer running it into one of three distinct parts (subprograms) in accordance with the salary currently being examined. Thus, this command tests a condition (income level), after which it branches to some distinct region of the program, depending on the answer. The repeat instruction enables the repeated execution of a certain part of the program. For example, if you want to compute the salaries of 100 employees, then there is no need to copy the subprogram responsible for this task 100 times. Rather, you maintain just one copy, to which you add the repeat command; this informs the computer that you wish to iterate through the computation 100 times. (There are also ways to indicate that you wish to process a *different* salary at each iteration, though in the same manner—using the same subprogram.)

Restricted as they are, computer languages give rise to an incredible richness of programs; for virtually any real-world problem, there is no single best program, and it is the programmer's job to do the best she can. To gain an inkling of the complexity involved, consider the following simple example: You want to write a program that prints twenty copies of "hello world" on the screen. One way to do this would be to write a sequence of twenty identical commands, print "hello world", where print is among the commands available in the computer-language repertoire (its function is simply to print the indicated text on the screen). A much terser program can be had, though, by making use of the repeat command: All you have to do is state that you wish the `print "hello world"` command to be repeated twenty times. No need to actually type twenty `print` instructions. Just type it once, and then type the magic word `repeat`.

Simple as this last example may be, it serves to convey the basic message: Writing computer programs is a complex task. Though the number of building blocks (commands) may be small, the number of ways that we can *combine* these blocks is huge. Think of the small number of brick types needed to produce the plethora of houses around us. This is just like a painter who has at her disposal but a few simple tools—brush, canvas, and some colors—out of which can emerge anything from a doodle to a Mona Lisa.

A computer programmer is a genie who grants computer wishes to *users*—the collective name referring to people who use computers. The genie—usually through the intermediary of a higher jinni known as "the boss"—is asked to grant a wish, say, imbue the computer with the ability to play checkers, which she proceeds to fulfill by translating the wish into a program. Alas, unlike her fellow lamp genies, the programmer genie cannot just snap her fingers to make the wish come true; rather she must work arduously for weeks, months, and sometime years on end in order to satisfy the users' desires.

But what if the computer programmer *could* just snap her fingers? Rather than labor strenuously to program the computer, what if the computer *itself* were to write the program? You're probably asking yourself how a computer program can just come to be like that. After all, I just said this was a very complex object, requiring the artistry of programmers. Let's take stock of what is involved: (1) a complex entity (program), (2) which is not to be designed by (human) hands, (3) but rather should somehow emerge within the computer environment. Forget for a moment that we're talking about computer programs. Does this scenario sound familiar? Complex entities that are not designed but rather emerge within some kind of environment. Yes, we're going to use our evolutionary-computing technique of the previous chapter to evolve computer programs. We already know how to apply evolution within the domain of engineering in general. But computer programs, being at the heart of all modern technology, deserve special attention.

In the world of computing, the field dealing with the evolution of computer programs goes by the name of *genetic programming*. Nichael Cramer, now a computing consultant, first proposed this idea in 1985, but it was John Koza from Stanford University who put genetic programming on the evolutionary-computation map, with the publication of his book in 1992: *Genetic Programming: On the Programming of Computers by Means of Natural Selection*. Before considering the evolution of computer programs, remember that our ultimate goal is to solve a problem (be it bridge building or finding a computer program). First off, we must design the artificial genome—the recipe that defines our evolving organisms, which are in fact candidate solutions to the problem at hand. We need also indicate the criterion by which the organisms shall be judged—we called this fitness. The computer can now take it from here. It first creates a population of hundreds (or even thousands) of completely random genomes, each representing a possible solution to the problem at hand. Each of these random individuals, collectively comprising the first generation, now takes the "survival" test, that is, it is assigned a score representing to what degree it satisfies the fitness criterion. Now comes the selection-crossover-mutation phase: The computer selects those individuals that will be allowed to reproduce, that is, to leave offspring in the next generation. Selection is carried out via a biased lottery in which fitter individuals stand a better chance of winning the right to reproduce. Having selected what gets to reproduce and how often, pairs of genomes come to exchange parts of their genetic material in the process known as crossover, followed by a few random mutations—small changes to the offsprings' genomes. Thus comes to be the second generation. Next, it's simply a matter of iterating the above process: evaluating the individuals of the new generation so that each one is assigned a fitness score, followed by the biased-lottery selection, crossover, and mutation, and so on, and on, and on.

The crux of genetic programming lies in the definition of the genome. With the general evolutionary-computing scenario of the previous chapter we did not restrict ourselves to any specific genomic description (as in our illustrative ten-brick genome, which consists of a simple list of ten spatial coordinates). With genetic programming our aim is to evolve computer programs; what, then, could be more natural than to maintain a population of programs? Thus, our genome becomes a computer program.

This sounds simple enough. You're looking for a computer program; then simply apply evolutionary computing to a population of programs, at the end of which—if all goes well—you'll end up with a good solution. This is similar to the bridge problem, except that in this case the final solution is a program. In reality, though, things aren't that simple. Let's return to our tax example. Suppose we're looking for a program that computes the amount of income tax to be paid by a taxpayer in our three-rung system (0, 20, or 30 percent). The computer sets out by creating the first generation of programs completely at random. This isn't hard to do. There is but a small number of possible commands to choose from, so the computer need only pick a random sequence of commands (which amounts to a random program). So far so good. Next, each program is to be evaluated and assigned a fitness score. This is not too hard. Each program can be run on the computer, "feeding" it with various sample solutions —taxpayer files along with the required tax amount. The fitness is the number of files for which the program computes the correct answer (the right amount of tax to be paid). What comes next? The biased-lottery selection. This, too, poses no problem. The computer cheerfully selects the lucky parents to mate and beget offspring into the next generation.

Selection is followed by crossover and mutation. In this case, crossover means exchanging parts between two parent programs, and mutation involves making random changes here and there to the resulting offspring. That's where we come up against a wall. Computer programs are *fragile*. If you start fiddling with them, randomly splitting, inserting, and removing pieces, then you'll end up with total "freaks"— programs that are illegal and cannot be run at all. (When you try to execute an illegal program, the computer "crashes"; it ceases everything and announces that a grave error has occurred.) With the ten-brick genome of the previous chapter, there was no problem; you always ended up with ten spatial coordinates, no matter what. With a program, though, you might end up with a total washout. For example, a perfectly good `add` command might turn into `abd` as a result of a mutation, a command that doesn't exist in the repertoire; or, due to crossover, a `print "hello world"` command might become `print add 2 and 5`, which is also illegal. The language understood by the computer posits that a print instruction must be followed exclusively by text and not by an add command. The fragility of computer programs stems from their strict rules of assembly, just as, say, with cars: If you begin to randomly mutate and exchange parts between automobiles, then you'll end up with a junkyard.

There is a prima facie solution to this problem: An illegal program can be declared unfit, that is, it gets assigned a fitness score of zero. However, crossover and mutation can be highly destructive; chances are that even with thousands of programs in the population, we'll end up with the vast majority of them having zero fitness. The selection process will then have very few options. We've lost the *variety* in the population, and—as noted in the previous chapter—variety is a sine qua non for evolution: Without it, evolution grinds to a halt.

Can this problem be solved? Can programs be rendered less fragile and thus more apt at undergoing evolution? "A fool sees not the same tree that a wise man sees," wrote William Blake in *The Marriage of Heaven and Hell*; and wisely turning to trees is the key to our solution. In most computer languages, the program is a sequence of commands—just as described above. However, there exists a less used though completely viable form for writing programs, known in computing parlance as *trees*. Consider the addition of two numbers, say, 2 and 3. The computer instruction that executes this operation is usually written as add 2 and 3 or simply 2+3. Now consider the following alternative form: a small "tree" with just a trunk and two branches—the trunk is labeled +, the left branch is labeled 2, and the right branch is labeled 3. Computing scientists like to draw simple treelike diagrams, in which the mathematical operations and numerical values appear in small circles that are connected by lines—the branches. The upper circle is known as the root, and—perhaps so as not to be too much of a Nature copycat—it is drawn at the top rather than at the bottom. Our 2+3 tree looks like this:

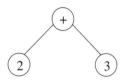

When wishing to avoid drawing diagrams, computing scientists use a textual notation for representing such trees. You write the trunk label to the left, followed by the branch labels from left to right, and enclose the whole thing within parentheses. The small 2+3 tree program would thus be written as (+ 2 3). This is in fact how the computer treats such trees: It doesn't deal directly with pictures but rather with these textual expressions.

Let's consider a slightly more complicated example. The following small program executes a numerical computation: Add 2 and 7, then subtract 3 from 5, and finally multiply the two results (since 2 + 7 = 9 and 5 − 3 = 2, the final result is 18). What does this program's tree look like? At the top is a trunk labeled × (for multiply), which ramifies into two thick branches, the left of which is labeled + and the right of which is labeled −; the left + branch further ramifies into two branches, labeled 2 and 7, and the right − branch also ramifies into two branches, labeled 5 and 3. The textual version is written as (× (+ 2 7) (− 5 3)). While this looks somewhat cryptic, there is a simple rule to help you decipher it: Each pair of parentheses contains exactly three parts—left, middle, and right—corresponding to the trunk, the left branch, and the right branch of the tree. The catch is that the middle and right parts (the branches) may be trees in their own right, rather than mere numbers. The ability to create elaborate programs is a result of this property; branches can themselves sprout offshoots. Here's the diagram that corresponds to the parenthetic expression:

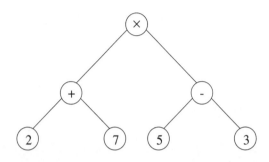

One can write highly complex programs using the tree format; in fact, they can get as complex as the classical sequence-of-instructions programs (there are tree equivalents of all the commands discussed above: add, subtract, read, print, conditional branch, repeat). Why have we gone to all this trouble? Because these tree programs are much more robust than the classical ones, able to withstand the winds of crossover and mutation. This is not a straightforward business. The genetic programmer still has to design the system with care. But, and this is the big advantage of the green approach, it turns out that trees can be rendered "evolvable" much more easily. *Crossover* involves the exchange of branches (possibly large ones with several offshoots) between two trees, while *mutation* involves small modifications to the branches themselves. For example, our (× (+ 2 7) (- 5 3)) program might exchange its (- 5 3) branch for a (+ 2 3) branch during crossover, resulting in (× (+ 2 7) (+ 2 3)). Mutation might then turn the 7 into a 5: $(\times \; (+ \; 2 \; 5) \; (+ \; 2 \; 3))$. The resulting tree diagram now looks like this:

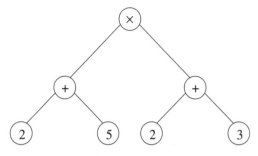

This genetically created program computes an entirely different value from that of its parent: the left branch $(+ \; 2 \; 5)$ amounts to 7, the right branch $(+ \; 2 \; 3)$ amounts to 5, and since the trunk is labeled $\times$ we have to multiply the two, giving a total of 35.

The setup is now ready; evolution can be turned loose. Our ingenious genetic programmer can sit back and let the computer run the process automatically, slowly but surely evolving better and better computer programs—lazy programming indeed.

During the past few years, John Koza (who wrote the first genetic-programming book) and his colleagues Forrest Bennett, David Andre, and Martin Keane have been applying genetic programming to the evolution of analog electrical circuits. If you look at a computer board (such as the memory add-on board you've just bought), you'll see that it consists of many black squares, the chips, connected by thin lines. Such a board is known as a printed circuit board, and it is a form of *digital* hardware. The "digital" part means that the hardware sees the world as a discrete collection of elements. The digital images we evolved in the previous chapter are composed of discrete pixels, each one of which can take on a color from a list of options (since there are a huge number of pixels, we don't "feel" the discreteness of the image unless we get very close to the screen). In fact, at its most fundamental, digital hardware deals with streams of only two symbols: 0 and 1. An evolved digital image resides in the computer's memory as a stream of 0s and 1s (which the computer can translate into the colorful picture that ultimately ends up on your screen). With *analog* hardware, the world is neither a dichotomy nor is it discrete, but rather it consists of a continuous spectrum. The quintessential example used to illustrate the difference between "digital" and "analog" is that of a watch: In a digital watch the seconds display advances in discrete, second-sized steps, while in an analog watch the seconds hand advances in a smooth, continuous fashion.

Most of the computer world is digital since such hardware is easier to design, easier to manipulate, and usually cheaper and more reliable. Analog hardware does, however, occupy some very important niches. Your sound system, for one, has various filters and amplifiers that are analog electrical devices. They resemble not what you see when you open up your computer, but rather what you'd see if you were to open up an old radio: a bunch of colorful wires, resistors, capacitors, and so forth. Whereas highly sophisticated computerized tools to aid digital designers do exist, analog-circuit design is somewhat of an art. The electrical engineers who design them are more like tinkerers who employ a large bag of tricks: the analog-circuitry lore that has accrued over the years. So why not employ the newest trick, evolution?

John Koza and his colleagues have used genetic programming to evolve such analog circuits. First, they defined the genome, in this case a tree genome that specifies an electrical circuit. The circuits are networks composed of a small number of basic components: wires, resistors, capacitors, inductors, diodes, transistors, switches, voltage sources, and current sources. Each serves a different purpose in the electrical circuit. We can draw an analogy between some of these components and a network of water pipes: Current is analogous to water; a current source is a water source; a wire is a pipe; a switch is a controllable dam; a capacitor is a water reservoir; a resistor is a water-flow controller (allowing more or less water to flow through it); and a diode is a one-way valve (allowing water to flow downstream but never upstream).

How does the tree genome, which is after all a computer program, represent such a network of pipes? It doesn't. The tree represents not the network itself, but rather an *ontogenetic* process that ultimately leads to the generation of the network. Ontogeny is the process by which multicellular organisms (such as you and your dog) develop through the successive divisions of a fertilized mother cell, known as the zygote. Starting from this single cell—a combination of your mother's egg cell and your father's sperm cell—*you* were created, with all your trillions and trillions of cells (we'll have more to say on ontogeny when we talk about embryonic electronics).

The tree genome Koza and his colleagues defined is a developmental program in the ontogenetic sense: A step-by-step procedure that, when applied to an "embryo circuit," results in a full-blown "mature circuit." In the water-and-pipes analogy, the embryo could be as simple as a pipe connected to a water reservoir. The tree genome contains a program that takes this embryo and develops it into an entire network. The genome contains instructions such as `add a component`, `delete a component`, and `change a component's value` (for example, have the resistor allow less water to flow through it). The mature organism (or electrical circuit) that emerges by running this developmental program on the embryo can be quite an elaborate network of pipes—or a complex electrical circuit.

Having defined the tree genome, our next step would be to determine how the genetic operators (crossover and mutation) work. Although more complex than the ones we saw above, the basic idea is the same. Finally, it remains to define the fitness, which depends of course on what we want to evolve: A network that is cheap (few expensive dams) while allowing as much water as possible to flow through it, or perhaps a network that occupies a very small area.

Going back to electrical circuits, one of the "creatures" Koza and his colleagues evolved was a woofer-tweeter filter, found in high-fidelity sound systems. A *tweeter* is a small loudspeaker responsive only to the higher acoustic frequencies and reproducing sounds of high pitch. A *woofer* is a loudspeaker, usually larger than a tweeter, responsive only to the lower acoustic frequencies, and used for reproducing sounds of low pitch. A woofer-tweeter filter sees to it that the low frequencies coming from your compact disk go to the woofer and the high frequencies go to the tweeter. Fitness in this case, was defined in terms of how good the individual circuit was in behaving as a woofer-tweeter.

Just as we evaluated the fitness of a Lego bridge in a virtual manner by having the computer examine its schematic plan, so too with the evolving electrical circuits. To test a circuit's fitness, you don't build it (which would be costly and time-consuming) but rather you use a program known as SPICE (simulation program with integrated circuit emphasis). This program, the de facto standard in the field of analog-circuit design, takes a circuit schema, simulates it in the computer, and outputs various electrical measures. These measures can then be used to compute the fitness score.

Here's a tree genome (using the textual notation involving parentheses) of an evolved electrical circuit. This evolved genome is a developmental program which, when applied to a simple embryo circuit, results in a circuit known as an amplifier. The evolved woofer-tweeter genomes are even larger (a few pages long) and more complex!

```
(ptposPR (ptposPR (ppR (ptposPR (ptposPR (ppR (ppR
(p2QNG_PNP_L )(dend )(p2QGG_PNP_L )
(p3Q_NPN_PL ))(dend )(p2QGG_PNP_L )(p3Q_NPN_PL ))
  (p3Q_NPN_NL )(res (num 19
19 1A 5 0 18 )(dend )))(p3Q_NPN_NL )(res (num 16 6
  1B 18 10
18 )(dend )))(dend )(res (num 1B C 7 18 1A
B )(dend ))(p3Q_NPN_PL ))(p3Q_NPN_NL )(res (num 1C
  1 7 6 0
18 )(dend )))(p3Q_NPN_NL )(res (num 1A 4 15 D 2 B )
  (dend )))
(ppR (dend )(dend )(cap (num D 3 12 8 E 12 )(dend ))
  (cap (num 1C E
E 1A F 1F )(ppR (dend )(p3Q_PNP_NL )(cap (num 1A 1B
  F 5 1
12 )(dend ))(cap (num 1C E E 1A F 1F )(cap (num 17
  1A 9 1C 4
18 )(ppR (dend )(cap (num 1A 1B F 5 1 12 )(dend ))
  (cap (num 1A 17
0 17 E B )(dend ))(cap (num 1A 1B F 5 1 12 )
  (cap (num 6 1B 14
1A F 1F )(dend )))))))))
(cap (num 16 D 5 6 1C 18 )(flip (p3Q_PNP_NL )))
```

So—is the programming profession in danger? Not quite yet. The programs evolved by the genetic-programming method are usually large and messy (just look at the typical evolved genome shown above). This is quite the opposite of human programming, which aims at being organized and terse. A human programmer, when asked to write a program, will attempt to produce a short, concise, and highly efficient piece. Genetic programming, on the other hand, produces "spaghetti" programs: huge trees with numerous weird offshoots and ramifications. It is extremely hard to make sense of them—in fact, it's rather akin to biologists trying to decode our own program, the human genome. As it turns out, when you delve into these evolved programs, you frequently find loads of "junk"—code that is of no use at all, and which could just as well be discarded. Again, this is quite similar to nature: Our genomes also contain junk code—unused portions of our DNA program. Both cases (genetic programming and natural evolution) involve the workings of a slow and tireless process that opportunistically latches onto any advantage, minute as it may be. Genetic junk can accumulate over the eons if it causes no harm, that is, no diminution in fitness, since evolution does not care to fiddle needlessly with the genome. And junk is something we do *not* usually want to see in computer programs. Finally, let's not be too quick in laying the programming profession to rest: We still need someone able to program this evolutionary scenario.

While the methodology presents its share of problems, genetic programming has been bustling with activity over the past few years. Researchers have been tackling fundamental questions in an attempt to render this technique more practical. Moreover, the method has been applied in a wide variety of domains, including market prediction, satellite control, robotics, and electrical-circuit design.

Lazy programming decidedly does not imply lazy programmers.

CHAPTER 3

Not Before You Do Your Homework

The most complex machine in existence today was designed not by humankind but by Nature: the human brain. Within a box roughly the size of a football—the cranium—Nature has managed to pack trillions of tiny processors (known as neurons) linked by quadrillions (thousands of trillions) of connections. However, the brain's "hardware" is not its most astonishing aspect—indeed, we shall probably be able to engineer such densely packed machines within the not-so-distant future (perhaps within a decade or two). It is the capacities this hardware gives rise to that leave us dumbfounded.

The mental feats of which our brains are capable are our pride and joy. Humankind celebrates its intellectual advances—be they in the arts, sciences, or technology—all of which are made possible by that "wetware" between our ears. Usually, when considering the pinnacle of the brain's capabilities, we tend to cite grandmaster chess players or mathematics prodigies. Impressive as their achievements may be, the brain is capable of performing even cleverer tasks, to which we seldom pay any attention. A 3-year-old child, any child, exhibits an *understanding* of the world that no computer to date has attained. The child has many abilities that outperform even the mightiest of current-day computers on several fronts. A few examples are "smooth" mobility in the world (without bumping into every obstacle), manipulation of objects (without destroying them—well, at least some of the time), and communication (language) skills. We're still struggling to build computers with these seemingly simple, though in fact highly elusive, capacities.

What is it that makes a 3-year-old child able to perform numerous tasks so much better than a modern supercomputer? There are several factors that come into play, not the least of which is the hardware—the densely packed brain. The hardware in itself, though, is insufficient to explain the brain's power; indeed, even when we get to the point of being able to stuff trillions of processors into a small box, we will still not be in possession of a brain. What is missing, then? What is at the heart of a 3-year-old's amazing capabilities but which still escapes computers? The crucial difference between the two is that a child can *learn*. Having talked about evolution, we now need to consider learning, another of Nature's fundamental processes that serves to enhance an organism's ability to adapt to its environment.

If we wish a computer to perform a certain task, then we have to feed it an appropriate program that contains precise instructions for doing the intended deed. A human being, on the other hand, does not require such preprogramming. Because we do come into the world with much innate knowledge (ultimately contained in our genome, the three billion-nucleotide chain of DNA), we are able to interact with our surroundings, modifying our behavior so as to gain a better understanding of the world and become more adept at surviving in it. This process is called learning.

There are several forms of learning—whether in children or in adults. One way is to have a mentor—such as a teacher or a parent—who can show us the ropes. Mentors are usually considered omniscient, not because they truly are so, but because their worldview is accepted unquestioningly ("the teacher always has the right answers"). We shall call this form of learning *supervised*. It is especially suitable for well-defined domains of knowledge, such as the science and math classes you took in school. The subject matter is precisely defined, and the teacher can go through it in an orderly manner, always in possession of the right answer, thus able to correct, or supervise, the student.

Another form of learning is what we shall call *reinforcement*, or learning by interaction. This occurs when the learner explores the world at large, being rewarded for good moves and punished for bad ones. Reinforcement learning need not necessarily involve a teacher or a parent—the environment itself can supply the feedback. For example, consider a child whose curiosity is aroused when she sees a boiling kettle on the stove, which she proceeds to touch. Punishment in this case is swiftly effected by the environment in the form of an induced pain reaction to the hot vessel. The child has just learned an important fact of life: A boiling kettle is definitely to be avoided. The environment can also be rewarding: When the child places three of his toy blocks one on top of the other, he is rewarded by an aesthetically pleasing tower. There is no need for a teacher or a parent to compliment the child; the sense of accomplishment and the delightful final result are feedback enough. The name "reinforcement learning" comes about because the learner is reinforced (whether positively or negatively) after taking an action.

Finally, there is a third form of learning known as *unsupervised learning*. Here, there is no teacher or any specific reward, and what is usually gleaned is not specific details about the world ("1 meter contains 100 centimeters," "avoid boiling kettles") but rather knowledge about relations and categories. Consider the following well-known example: A child is told that the four-legged, furry animal playing on the front lawn is a "dog." A few days later, during a trip to the countryside, the child sees a four-legged animal and immediately cries out with pleasure: "There's a dog." In fact, the animal in question is a horse, but the child placed it within the dog category due to the obvious similarities (four legs, head up front, tail in the back). While there may be a teacher or a parent to provide the correct answers (dog, horse), these are but arbitrary linguistic tags. What matters here is the emergence of such categories of knowledge. We might imagine the child growing up alone on a desert island; she would still exhibit this form of unsupervised learning. Though obviously not arriving at the correct English word, the child will nonetheless be able to categorize the island's fauna, slowly refining these categories as she grows up. For example, dog and horse might first belong to the same class: four-legged, head up front, tail in the back; later, as the child matures, she will separate them into two classes: horse (large, used for riding) and dog (small, a human's best friend). Unsupervised learning thus serves to form classes and to discover relationships about the world around us. (Note that while there may be classifications that are more useful than others, there are no a priori correct classifications: Our island child might very well place both dog and horse in the same class—that of friendly animals, for example.)

Learning enables us humans to come into the world only "partially programmed," to then continue our education throughout our lives. This is in stark contrast to computers, which usually need to be given the full program in advance. Perhaps the greatest advantage of the learning process is that it tends to go hand in hand with what is known as *generalization*. When we learn some property of the world, we do not need to be presented with *all* possible instances of it, only with a (usually small) number of examples. One boiling kettle is enough. Even when a new, completely different-looking kettle comes along, the child will still be wary of it. We recognize with ease a new horse as such even though we've seen only a few examples before. This is due to our ability to generalize, that is, to correctly interpret novel situations based on previously gleaned knowledge. Though we frequently do so with ease—hardly giving it any thought at all—some cases may challenge us: Is that three-wheel vehicle a car or a motorcycle? It has handlebars so it's more of a bike, but what if it had a steering wheel? Generalization is not always straightforward.

The ability to learn is so potent that it seems natural to ask whether machines can be made to do so: Can computers learn? An affirmative answer was given by researchers in the 1940s and the 1950s, whose collective efforts gave rise to the field known as *artificial neural networks*. Their basic approach consisted of drawing inspiration from the way biological neural networks work (for example, the human brain) in order to imbue computers with the ability to learn.

The field of artificial neural networks has known its ups and downs. In 1943, Warren McCulloch and Walter Pitts from the University of Illinois published a paper entitled "A Logical Calculus of the Ideas Immanent in Nervous Activity," considered by most to be the official launcher of the field. During the 1950s, when research into *artificial intelligence (AI)* began, there was a flurry of interest in artificial neural networks. One of the most celebrated works was that of Frank Rosenblatt, an American psychologist who proposed in 1958 the *perceptron*, which was a more sophisticated neural-network model than the one proposed by McCulloch and Pitts. For about a decade there was much excitement with this newborn, brain-inspired form of computing. Then, in 1969, Marvin Minsky and Seymour Papert from the Massachusetts Institute of Technology published a book in which they showed that perceptronlike models had certain fundamental limitations and were unable to tackle important classes of hard problems. Though not their intention, it seems that this book was a major cause of artificial neural networks' falling into disfavor during the 1970s, both among academics as well as with funding agencies (especially in the United States).

In 1982 the American physicist John Hopfield proposed a completely different kind of artificial neural network that caused much excitement and marked the rekindling of the field's flame. This revival was clinched with the publication of a seminal two-volume work by the PDP (Parallel Distributed Processing) research group at the University of California, San Diego headed by David Rumelhart and James McClelland. They showed, among other things, how the limitations presented by Minsky and Papert could be overcome. (It later turned out that part of their work was actually a rediscovery of earlier work done by Paul Werbos as a doctoral student at Harvard University; such rediscovery in scientific research happens from time to time.)

Over the past 15 years, the field of artificial neural networks has come into its own, with such networks applied these days not only by academics but by industrial firms, performing such tasks as recognizing faces and predicting the future of financial markets. Computers can learn (and generalize). They do so not by being programmed in advance but by being presented with examples of the problem at hand, which they must solve correctly—that is, they must do their homework.

Before proceeding to explain how artificial neural networks function, let's first take a crash course on the brain. As noted earlier, the human brain contains some trillion-odd tiny processors known as neurons—the fundamental functional units of nervous tissue. The neurons are connected among themselves, forming a (neural) network with some quadrillion-odd wires. The brain's functioning is based on the transmission of information in the form of electric impulses along these wires. The basic unit of computation—the neuron—has three main parts: the dendrites, the nucleus, and the axon (see Figure 3-1). The dendrites are arranged in a treelike manner, known as the *dendritic tree*, which conveys incoming signals (from other neurons) to the nucleus. To picture this, think of a tree whose thick lower branches are connected to the nucleus (the trunk), and the leaves are connected to other neurons. Signals arrive through the leaves, onto the branches, and into the nucleus. The nucleus pro-

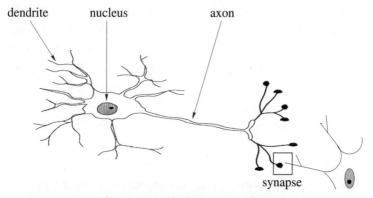

Figure 3-1   The neuron, the basic unit of computation

cesses the incoming dendritic signals (the input) and sends them (the output) along a single wire: the axon (in our tree image this would amount to a single root stem).

The brain's power lies not in the computation performed by a single neuron but rather in the collective computation performed by the network as a whole. Having processed the incoming signals, the nucleus sends an output signal along the axon; at a certain point this signal branches out, forming connections with several dendritic leaves, which belong to different neurons. This is at the heart of neuronal communication (and computation): Every neuron receives input signals from numerous other neurons (through the leaves of the dendritic tree), processes them, and then transmits an output signal through its axon to many other neurons.

The point where an axon meets a leaf of another neuron's dendritic tree is known as a *synapse*. It is here that the electric signals traveling along the axon get "bogged down." The synapse is a chemical element, and the signals traverse it at far slower chemical speeds (think of a jet plane crossing a water barrier by resting on the deck of a giant sailboat). The synapse does not act as a mere relay but may cause the passing signal to change; moreover, the synapse itself—or rather its chemical properties—may change in time. The synaptic changes are thought to be of crucial importance to learning. This is even more pronounced where artificial neural networks are concerned. In this case learning usually takes place by making changes to the artificial synapses—the points of contact between two neurons.

It is instructive at this point to consider the differences between brains and computers. The brain operates by employing a huge number of tiny, simple processors, all of which work in parallel. The communication among processors is quite slow because of the chemical barriers. A standard computer, on the other hand, contains one complex processor, which does one thing at a time (this is known as a sequential machine—as opposed to a parallel one, such as the brain). Since everything is electronic (no chemical elements), the computer is much faster—at least a million times so— than the brain. Nonetheless, the brain's massive parallelism more than compensates for its sluggish elements, and it is thus able to outperform the serially operating computer on several fronts. Another interesting difference involves the way information is handled. In a computer one piece of data (say, your best friend's telephone number) is stored in a precise location in memory (somewhere on the hard disk). If this location is destroyed, then hopefully you've got the number memorized in your brain, which stores information in a distributed fashion. That is, the phone number is "spread over" many neurons and synapses, so losing a few of those won't do any harm; you'll still be able to call up your friend. (In fact, from about the age of 20 onward, we lose a few thousand neurons every day, which we seem to handle quite well.)

What I've outlined above is but a rough sketch of biological neural networks, the reality being vastly more complex. Even at the neuron level, biologists have found that the processing taking place is not as simple as it was once believed. Let's not forget that we're talking about an extremely complex machine, about which the extent of our knowledge is but a fraction of the extent of our ignorance. There are still many fascinating unanswered questions about the brain: What neuronal processes take place that allow you to recognize your grandmother? How is language processed (and why are we the only earthly animal to exhibit such developed linguistic skills)? What is consciousness, and how does it arise? These are but mere samples of the many mysteries still present in brain research. Where artificial neural networks are concerned, however, this lack of knowledge does not pose a major problem. Most such networks, while inspired by the brain's basic mechanisms, are quite far removed from the actual biological details. The goal is not to model the brain but rather to be able to create learning computers. (A related but separate field of study, known as *computational neuroscience*, uses neural networks as a mathematical tool for explaining the brain's functioning; researchers in this area *do* seek to remain as faithful as possible to the biological reality.)

An artificial neural network is a network of *artificial neurons* (Figure 3-2). Whereas the biological neuron may

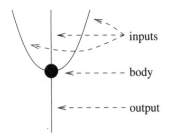

Figure 3-2   An artificial neuron

have turned out to be more complex than previously assumed by neuroscientists, the artificial neuron is quite a simple processor. As with its biological counterpart, such a neuron is made up of three parts: dendrites, a nucleus, and an axon. Often, we don't even use the biological terms but rather talk of the input, the neuron body (or processor), and the output. You can picture the artificial neuron as a multipronged fork: The several prongs correspond to the input lines (dendrites), the base corresponds to the neuronal body (nucleus), and the handle corresponds to the (single) output line (axon). As you can see, the input lines are arranged in a much simpler manner than the biological dendritic tree: a row of prongs that directly enters the neuron body.

Thus far we have considered the artificial neuron's structure; now it's time to talk about the computation it performs. Each input line is designed to hold but a simple number at a given moment, which is the value sent to it by another neuron. Furthermore, each such line is also associated with a so-called synaptic weight—a number that embodies the changes that the synapse makes to passing signals. The neuron now computes a *weighted sum* of its inputs and its synaptic weights. What is a weighted sum? Suppose you're in college taking an introductory course on computing. The professor informs you at the beginning of the semester that the final grade will consist of the average of three elements: a final exam, a project, and a class presentation. Being a good student, you score a 9 on the final exam, a 10 on the project (way to go!), and an 8 on your presentation (need to polish those oration skills). What is your final grade? You simply add up all three marks ($9 + 10 + 8 = 27$) and divide the result by 3. Congratulations, your final grade is 9. This calculation is (of course) quite familiar; it's called computing an average. Another way of doing this is by considering each element to have a "weight" of one-third, and proceeding to compute a weighted sum: 9 multiplied by $\frac{1}{3}$, plus 10 multiplied by $\frac{1}{3}$, plus 8 multiplied by $\frac{1}{3}$; not surprisingly the result is the same (9). Now let's assume that the three grade constituents are not equally weighted: the exam represents 30 percent of the final grade, the project represents 60 percent, and the presentation accounts for only 10 percent. The final grade is again given by the weighted sum of the three—30 percent of 9 (2.7), plus 60 percent of 10 (6), plus 10 percent of 8 (0.8)—giving a total score of 9.5. Great! You've just moved from an A to an A+, and all it took was (the professor) fiddling a bit with the weights.

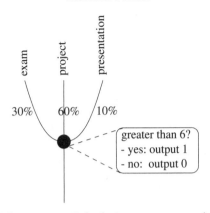

Figure 3-3   Calculating a course grade
using the artifical neuron

Figure 3-3 shows how the artificial, three-pronged neuron calculates the course grade. Each line holds one of the
three grade constituents, with the respective synaptic weight
set to mirror the professor's choice (the simple average can be
attained by setting all weights to one-third, and the second,
project-biased grade can be attained by setting the weights to
30, 60, and 10 percent). We're almost done explaining the
artificial neuron's computation—we need just add one small
detail. Usually, the weighted sum does not pass directly to
the output axon (to then be communicated to other neurons);
rather, this computed sum first passes through a *threshold
barrier*. For example, one such simple barrier is the following: If the computed grade is greater than 6, then output 1;
otherwise, output 0. In this case we might consider that the
threshold represents a pass/fail barrier. To summarize, an
artificial neuron performs but a simple calculation: the so-
called weighted-sum-and-threshold. (There are several types
of artificial neurons, most of which are variations on what
I've just outlined.)

Now that we know what an artificial neuron looks like (a multipronged fork) and what it does (weighted-sum-and-threshold), it's time to build a network (as with the single neuron, artificial networks of neurons are also far simpler than their biological counterparts). We first align several forks (neurons) in a row with the prongs pointing upward; this is called a *layer*. Next, we add another row beneath the first, and connect wires between each handle (output or axon) of the first-layer forks, and each of the prongs (inputs or dendrites) of the second-layer forks (since we're talking forks here, you can think of the wires as spaghetti strands). This can then go on for a few more layers—always connecting the outputs (handles) of the top layer to the inputs (prongs) of the layer underneath (in practice one rarely needs more than four or five layers).

This structure, shown in Figure 3-4, is referred to as the network *architecture* or *topology*. The structure just described,

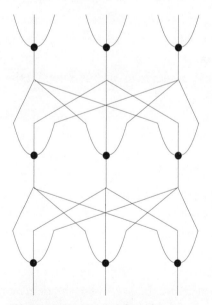

Figure 3-4   Network architecture

known as a *feed-forward network*, is but one possible topology of many (though a commonly used one at that).

How does the network function? Let's suppose that there are three layers, each with ten neurons and that the network's task is to predict tomorrow's value of a certain stock. The top row is known as the input layer. We present it with what we know about the problem, in this case the stock values for the past ten days. (To be more precise, each neuron has ten input lines: The left input line receives today's stock value, the one to its right receives yesterday's value, and so on until the rightmost input line receives the stock value of nine days ago). Each input-layer neuron now calculates—in accordance with its synapses—the weighted-sum-and-threshold, resulting in an output value. Because every one of these neurons has wires (axons) that connect it to all neurons of the second layer, it can now send along these wires the output value it has just computed. Looking at this from the point of view of a second-layer neuron, it has just received on its ten input lines ten computation results—one from each of the input-layer neurons. It can now compute—in the same weighted-sum-and-threshold manner—its own output, which it then sends to the third-layer neurons. These latter compose the last layer, also known as the *output layer*; their output axons are left "dangling" since there is no further computation. We—the network users—can now examine these axons, which contain the network's final result. For example, we might design the network such that the leftmost neuron of the output layer holds the answer: tomorrow's stock value. Thus, our forks-and-spaghetti network comes to compute—bon appétit!

The network can be seen to exhibit "waves" of computation. All neurons of the first (input) layer do their bidding, send their results to the second-layer neurons, which then compute and send their results to the third-layer neurons, whose resulting computation is the final answer. How can anything useful come out of this process? After all, we've seen how simple each neuron is, and all we've done here is stick a few dozen of them together and let them do their neuronal dance. But that's exactly the secret of artificial neural networks: The power lies not in the computation performed by a single neuron but rather in the collective computation performed by the network as a whole. Since each neuron has different synaptic weights, it will compute an output that is different from those of its neighbors, which it then transmits to the next layer; the neurons of this layer, upon "seeing" different incoming signals, will in turn compute different outputs.

How do these synaptic weights come to be? They are *learned*. You do not program these numbers by hand, but rather let the artificial neural network learn them by itself. As with humans, there are three principal learning methods: supervised, reinforcement, and unsupervised (earlier I used these terms to describe human learning, though in fact they are drawn not from the field of human psychology but rather from the domain of artificial neural networks). There are several methods known today for *training* a network by providing it with samples of the problem at hand. In the case of supervised learning, the network is given sample queries along with the correct answers; for example, in order to recognize faces, the network will be fed a multitude of sample pictures along with the correct name tags ("Warren McCulloch," "Walter Pitts," and so on). Reinforcement learning involves environmental feedback; for example, training such a network to play backgammon can be achieved by having it play various human players, rewarding it when it makes good moves and punishing it when it makes bad ones. With unsupervised learning, we need only present samples, for example, several pictures of dogs and horses, such that the network comes to differentiate between these two concepts.

Usually, we distinguish between two operational phases: the training (or learning) phase and the test (or evaluation) phase. During learning the network is presented with sample problem instances, and the synaptic weights are constantly modified. We set out by assigning synaptic weights completely at random. Next, each sample input is presented to the network, which works its magic, executing the neuronal waves of computation to produce the final answer. It is at this point that the synaptic weights are modified; the exact way in which this is done depends on the specific learning methodology we're using. Dozens of such methodologies—known as *learning algorithms*—are in existence today.

The idea underlying learning algorithms is quite simple (though the details may be less so). When the correct answer is known in advance (as in supervised learning) or when we at least have an idea of what good and bad moves are (as with reinforcement learning), then, we do the following after each sample is run through the network. If the difference between what the network produced (called the *actual output*) and what should be produced (called the *target output*) is large, we make small modifications to the synaptic weights so as to reduce this difference (the exact way in which this is done is what gives rise to the plethora of learning algorithms). Thus, the next time this input (or a similar one) comes along, the network's response will be somewhat better (if it is not—then the synaptic weights are again modified in the same manner). With unsupervised learning, things are a bit different since no a priori correct answer is available. The idea here is to let the network run through many samples, for example, several pictures of dogs and horses. Since they are different, pictures of dogs will produce different neuronal waves of computation from pictures of horses. The learning algorithm then works to enhance these differences (again, through modification of synaptic weights). Thus, the two concepts—dog and horse—become well separated into two distinct classes.

Once we've finished training the network, we pass on to the test phase: Synapses are no longer modified, and the network is tested on samples that it has never seen before (but which belong, of course, to the problem at hand). Thus, we might test the network on new facial pictures, novel stock values, or a new backgammon player. This last phase serves to evaluate to what degree the network can generalize (remember that this refers to how well the network can correctly interpret novel situations based on previously learned knowledge).

Artificial neural networks have been applied extensively to perform *pattern recognition and classification.* This term encompasses a wide range of problems, which all have the identification of features and motifs within a given object in common: faces in a photo, airplanes in a radar image, submarines in a sonar scan, your mom's voice in a recording of a cocktail-party conversation, and so forth. Take handwriting recognition, for example, a problem which is often hard even for us (I for one belong to the "I can't read my own handwriting" group.) A computer able to read handwritten documents would have numerous applications, from comprehending addresses on envelopes to automatically transforming your squiggles into an electronic document. This problem has re-ceived much attention over the past few years, and there now exist standard databases that are used by researchers working on handwriting recognition. Such databases contain samples of handwritten letters and digits from thousands of different persons and can be used in supervised learning. For each sample letter (a handwritten B), there is a known correct answer (the letter B). You can load one of these databases from the Internet and pit your newly proposed artificial neural network against the best handwriting recognizers on the market.

If artificial neural networks are such good learners, then why aren't they everywhere? After all, isn't it much easier to solve a problem by providing examples rather than by writing a full-blown program? Yes and no. The last statement is indeed true—for certain types of problems, such as those mentioned above. They are characterized by our ability to represent them in a form that can be presented to an artificial neural network, as well as by the ability to generate sample instances (this is not always as simple as it sounds). There are, however, many problems that we know very well how to solve using classical programming techniques, such as multiplying large numbers or searching for a given name in a phone book; there is no need to resort to learning in such cases. Artificial neural networks (as well as other methods described in this book) become candidate solutions for problems on which standard computers perform poorly; interestingly, many of these are exactly those tasks which we humans are good at (such as recognizing faces). This is perhaps the major reason artificial neural networks have not replaced your desktop computer—they're not good at *everything*, just at certain (hard) tasks. There are also other issues of a more minor nature; for example, the networks don't always succeed in learning the task at hand (alas—as in school—some students fail), and even when they do, training may take a long time (days, weeks, and sometimes longer).

Artificial neural networks are based loosely on the workings of the brain, which, from time to time, has been known to *think*. So can artificial neural networks think? This question is in fact part of a broader one, that of whether computing machines in general can think. While we now know how an artificial neural network functions, one thing I haven't told you so far is how to *implement* one. In the brain, the neurons and the synapses are implemented in "wetware," using biological, carbon-based material. Artificial neural networks are, for the most part, implemented as *simulations on standard computers*, such as your PC. (There are researchers working on hardware implementations using brainlike electric circuits or malleable chips such as the configurable processor we will encounter later on; these works are still, however, a minority within the field.) If an artificial neural network is simulated on a PC, then the question of whether it can think translates into the broader question: Can computers think?

Not yet. Very few artificial-intelligence researchers would dare proclaim that we are currently in possession of thinking machines. This question has a centuries-long history and has been the cause of much debate over the past 50 years, during the modern era of computing. "To ask the hard question is simple," wrote W. H. Auden, but to answer it, well, that's quite another story. We consider the issue of intelligence in an evolutionary context later on in the book. For now, I wish to make one thing clear: Artificial neural networks *cannot* think (that's what most experts think), but they *can* learn. They are not as good learners as we are, but they are making progress.

We have thus managed to imbue computers not only with the ability to evolve but also with the ability to learn. In the next chapter we see how researchers in the field of adaptive robotics combine these two processes, evolving artificial neural networks in robots.

Since computers can learn, we might well wish to consider what we should teach them. This brings to my mind a scene from the motion picture *Terminator 2: Judgment Day*.

While recuperating in a run-down service station, the following conversation takes place with young John Connor, his warrior mother Sarah, and the humanoid machine known as the Terminator:

JOHN: Can you learn? So you can be … you know. More human. Not such a dork all the time.

TERMINATOR: My CPU is a neural-net processor … a learning computer. But Skynet presets the switch to "read-only" when we are sent out alone.

SARAH: Doesn't want you thinking too much, huh?

TERMINATOR: No.

JOHN: Can we reset the switch?

CHAPTER 4

D 'N' A: Darwin and Asimov

Even before the word "robot" was coined in the 1920s (etymologically originating in the Czech word *robota*, meaning work), humanity had always been fascinated with the idea of creating a machine in its own image. During the second half of this century, "robot" has become a household word owing in large part to the celebrated science fiction writer Isaac Asimov. I still remember my first encounter as a teenager with the three laws of robotics:

1. A robot may not injure a human being, or, through inaction, allow a human being to come to harm.
2. A robot must obey the orders given it by human beings except when such orders would conflict with the First Law.
3. A robot must protect its own existence as long as such protection does not conflict with the First or Second Law.

And, of course, is there anybody (at least on this planet) who doesn't know who R2D2 and C3PO are?

But science fiction is one thing, while science is quite another. One of the starkest unfulfilled promises of computing science and engineering to date has to do with *robotics*: the technology dealing with the design, construction, and operation of robots. (Incidentally, the introduction of the term *robotics* is attributed by the *Oxford English Dictionary*, 2d ed., to a 1941 science fiction story by Asimov.) Extensive research efforts in the field of robotics began in the 1950s, both by engineers, who were interested in the formidable technical obstacles that need to be surmounted in order to construct a real-world robot, as well as by computing scientists, who sought to imbue such machines with intelligence. The task turned out to be enormously difficult, and, about 50 years later, we still have a long way to go.

I'd like to clarify at this point what I mean by "robot," since this term is at the heart of two quite separate domains. First, there are the *industrial* robots, such as those one finds in virtually every modern-day factory, whether fabricating cars, computer chips, or T-shirts. Such robots exhibit hardly any semblance of humanity: They're bulky machines, generally riveted to the floor, jerkily going through a fixed repertoire of movements in a tireless fashion. Industrial robots, though, have been highly successful and are very well understood today by engineers. We know how to design them, how to build them, and how to maintain them—in short, how to make them tick.

The other kind, the robots that usually capture our imagination, are by contrast those with which this chapter is concerned and generally go by the name of *autonomous* robots. First and foremost they are characterized by their ability to *move* in the world at large and to *interact* with it. The typical extant mobile robot looks more like R2D2 than like C3PO: a body of metal and plastic, strewn with various protruding appendages, which propels itself either on wheels or on a set of spidery legs. (Bipeds are much harder to balance, while several legs, arranged in a spiderlike fashion, are decidedly easier.)

These appendages are of two types, *sensors* and *actuators*; they enable the autonomous robot to interact with its surroundings. The sensors act as the robot's input mechanisms— its "eyes" and "ears"—allowing it to sense its environment. They need not be limited to the spectrum of human sensations: The eyes might be tuned to the infrared spectrum, while the ears might hear ultrasound. The actuators are the robot's output mechanisms—its arms and legs—enabling it to control its own movement as well as to make changes to the environment. Again, these appendages may be nothing like human ones, with wheels instead of legs and mechanical grippers rather than hands.

The ability to move, sense, and act upon its surroundings is at the heart of the robot's claim for autonomy. The central goal of research in autonomous robotics is to construct robots that can get along without us. We want to be able to turn them loose and have them carry out their intended function without the need for constant supervision.

And just what is their intended function? In *The Notebooks of Lazarus Long* Robert A. Heinlein wrote: "A human being should be able to change a diaper, plan an invasion, butcher a hog, conn a ship, design a building, write a sonnet, balance accounts, build a wall, set a bone, comfort the dying, take orders, give orders, cooperate, act alone, solve equations, analyze a new problem, pitch manure, program a computer, cook a tasty meal, fight efficiently, die gallantly. Specialization is for insects." We humans are definitely quite versatile creatures, able to perform an astonishing range of tasks. Yet even an insect, lowly as it may seem, does amazing things: It is able to change direction, plan a route, butcher an enemy, conn a piece of food, design a retreat, balance its body, build a nest, cooperate, act alone, solve problems, analyze a new problem, pitch manure, find a tasty meal, fight efficiently, die gallantly. Impressive, no?

Extant autonomous robots are far behind even the insect level. A small mobile machine able to exhibit the versatility of your average cockroach would make any robotics researcher drool. Why are such robots so hard to come by—especially in view of the enormous success their relatives, the industrial robots, have been enjoying? Let's consider a typical industrial robot: a mechanical arm riveted to the floor of some T-shirt factory, whose function in the production line is to stitch the three buttons of the V neck. First, note that this robot isn't mobile (it has no legs or wheels)—and moving is hard! Look how long it takes a human child to learn to be mobile. Moreover, the arm's task is simple and precisely defined: Place three buttons at three predetermined positions, and stitch them in a prescribed manner. And, to top it all, the environment is completely fixed from the outset and does not change by a single bit as the robot does its work: The T-shirts arrive in an orderly manner, each placed in exactly the same position as the previous one. This renders the programming of such machines a doable task using current technology. For example, one way of doing this is by walking the mechanical arm through the sequence of movements it needs to stitch the three buttons. This sequence is thus programmed once at the outset and does not need to be changed afterward (unless we decide to modify the specifications: Place only two buttons, stitch pants, and so on).

The most important difference between industrial and autonomous robots has to do with their *environment*. As we've just seen, the industrial robot has an easy life. Everything is handed to it on a platter, and, what's more, it's exactly the same platter all the time, the environment being *static*. An autonomous robot, however, must deal with a *dynamic* environment, and a much more complex one at that. Indeed, our ultimate goal is for such critters to be able to carry themselves with grace in *our* world.

To see just how complex our world is, consider a robotics consumer product I'm sure we'd all rush to buy were it available on the market: Dr. Clean—aka Dean—the autonomous housemaid. Dean cleans, arranges, vacuums, and tidies; in short, it keeps your house spotless and shipshape; and it does so all on its own (no need for you to guide it the way you do with your average vacuum cleaner). Just like Rusty, the family dog, it has its own little corner in your house where it "sleeps" (turning itself off when not bustling about) and "eats" (recharging its battery).

Dean loves those quiet suburban mornings when the family has left for work or school and it has the house all to itself. It opens its eight eyes and starts to roam around the house. (Why eight eyes? Actually, the question is, Why only two? With eight eyes, Dean can cover not only its front, but also its sides and back at the same time.) Whenever it sees a piece of trash, Dean picks it up and places it in an inner container within what passes for its belly. When full, Dean ambles over to the kitchen and negotiates a nice profitable transaction with Mr. TrashCan, at the end of which the latter is full and content while Dean is again light as a feather. Speaking of the kitchen, the leftovers from breakfast are still on the table, but they don't stand a chance against Dean. Fifteen minutes later, you could eat off the floor. Dust on the TV? Not a problem: Dean delicately wipes it off with its mechanical gripper. After climbing up to the second floor, it enters the children's room—ugh, what a mess! The place looks like the aftermath of some horrible war, and, what's worse, it seems to be the quarters of the *losing* side. It takes Dean a full half hour to set everything straight: rearrange the toys, fold and place the clothes strewn about back in the closet, clean the carpet, and of course dust the entire room. Upon completing a final tour of the house, making sure that everything is OK, Dean's internal battery monitor warns him that supplies are low. Time to go back to its corner for a well-earned meal and some sleep.

The tasks performed by Dean seem quite trivial and unintelligent to us: Just your run-of-the-mill house cleaning; we do it all the time with ease if not without complaint. You'd be surprised to learn, though, just how demanding these deceptively simple tasks are and how difficult it is to build such an autonomous housemaid. First off, Dean has to be able to walk, or, to be less anthropomorphic, to move; and it must do so without bumping into every single object in its way. We've already noted how hard this is. As adults we are able to walk around effortlessly and unmindfully, perhaps even carrying on a conversation at the same time. Nonetheless we spent the first two to three *years* of our lives learning to master the technique. And the learning process could be quite painful at times, with surprises popping up at each step of the way. ("Ouch, where did that table come from?") While it is true that many animals can walk within minutes of their birth (newborn deer, for example), this does not diminish the task's difficulty. It just means that this behavior had become incorporated within the genome during the evolutionary process. (Usually there is a good reason: Many of these animals must face dangerous predators as soon as they enter the world, and the ability to walk straightaway is thus a boon to survival.)

Moving about in a cluttered household is something that we cannot preprogram Dean to do by walking it through the motions—as we did with the T-shirt robotic arm. The sequence is too complicated, and, more importantly, the house is a dynamic setting. Things are *never* in exactly the same place (even in the home of an obsessive-compulsive person). Besides the behavioral aspect of moving, namely, how to do so, there are also many technical questions concerning the fabrication of a mobile machine. Should it have wheels (which are good for smooth terrains, such as a large parquet floor), legs (which are better where rough terrains are concerned, such as a typical house—with thresholds, carpets, and what have you), or perhaps both? How does it climb stairs? What kind of control mechanism must it have in order to be able to stop in time (before the "ouch")? Moving is decidedly hard.

As difficult as this may be, getting Dean to move about smoothly—without bumping into obstacles—is but the first step (both literally and figuratively). The second—and much more arduous step—is for Dean to be able to sense its world (the home) and make changes to it (remove trash, dust shelves, arrange clothes). Again, while deceptively simple, this is in fact a huge problem since it requires knowing many things about the world. Numerous things that seem trivial to us—having learned them throughout our lives—are not at all clear to Dean: Carpets are not obstacles (since you can step on them) yet TVs are; yesterday's paper is garbage, but the book lying on the floor is not; gripping a glass must be done with much more care than gripping a football; folding the clothes strewn about the floor is definitely a good idea, while attempting to fold Rusty is probably somewhat more risky; and the list goes on and on. The ability to sense the world, to make a correct decision, and then to follow up on it still remains—

after years of research—an elusive goal. And this holds even if we limit the environment to that of a house interior and the task to that of cleaning (Dean is not expected to hold up its end in a discussion about the upcoming elections).

As with mobility, sensing and effecting actions also involve several fabrication problems. We humans come into the world equipped with superb sensors—such as ears and eyes—as well as highly flexible and versatile actuators—such as arms and legs. We're not yet able to mimic Nature's achievements by furnishing robots with such organs. We can build machines with more raw power, like cranes (which are stronger than us) and cars (which are faster), but we cannot attain Nature's finesse. The human hand can handle a wide range of materials—whether bricks, glasses, feathers, or babies—without causing them even the slightest damage; existing robotic arms are much rougher.

Yet even if (or perhaps I should say when) we are able to build Dean and equip it with the basic abilities to move about, sense, and react, there will still be something missing—an even more difficult function, problem solving. It is not so much solving mathematical equations or the sensing and actuating problems we've just discussed. Rather, I'm referring to a different form of reasoning, which involves the numerous small questions that Dean must answer when cleaning the house: Does a locked door mean it shouldn't enter? Perhaps it should use the key or call the owner? What happens if it runs into an unknown person roaming the house (friend, Aunt Matilda; or foe, cat burglar)? What must be done in the case of a fire? How sloppy can this family get, leaving all those hundreds of small colorful pieces on the living-room floor? (Oops, the beautiful puzzle you've all been working on till 4 o'clock in the morning has just been trashed.) These questions all share one thing in common:

They involve novel situations that Dean has not encountered before, but need nonetheless to be addressed and resolved. This level of functioning is perhaps somewhat less crucial since we can always decide that when it comes across such a problem, Dean responds in a preordained way, for example, by returning to its corner or by calling up the owner.

Building autonomous mobile robots is far from easy. I'm sure most of us would like to own such a Dean, which would relentlessly maintain our castle in a state that would put the mother-in-law institution out of business. You'll be happy to learn that engineers and computing scientists have not been idling about. The past 40 years have seen intensive research efforts aimed at building robots that can carry themselves with at least some modicum of autonomy. Let's see what has resulted from all this hustle and bustle. (I'll concentrate on the behavioral side of things, leaving out the fabrication issues, though these also admit many fascinating tales.)

Up until a few years ago the main approach was that of attempting to imbue robots with as much knowledge as possible about the world in general, and about their mission in particular. You wanted to instill the machine, a priori, with the facts of life, so that it could then go out and do your bidding. We have seen, however, what a hefty sack of wisdom is needed in order to do even so-called simple stuff. Over the past few years a new direction has emerged that goes by the name of *adaptive robotics*.

Let's go back to children: They come into the world "equipped" with many helpful devices, including sensors, actuators, and a brain, which have evolved over eons. They then complement their innate Nature with nurture, learning more and more about the world as they grow up. These processes underlie evolutionary computation and artificial neural networks. Rather than try to instill robots with everything they need to know, with every bit of their behavioral repertoire, why not set out with very simple machines and have them evolve and learn? "Why not indeed?" responded several researchers, giving rise to the field of adaptive robotics.

Suppose you want your robot to be able to move about in the office without bumping into chairs, tables, and whatnot; in robotics this is called *obstacle avoidance*, a behavior whose attainment has become a benchmark problem among adaptive roboticists. An actual real-life robot, used by hundreds of research laboratories worldwide, goes by the name of *Khepera* (named after a divine scarab beetle which was the dawn manifestation of the Egyptian sun god) (see Figure 4-1). Designed by Edoardo Franzi, Francesco Mondada, and André Guignard from the Swiss Federal Institute of Technology in

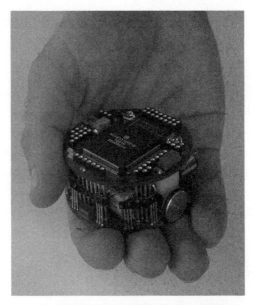

Figure 4-1 The Khepera
COURTESY OF ANDRÉ ("CHICO") BADERTSCHER

Lausanne, the Khepera is a small, saucerlike critter about the size of your average mouse (roughly 6 centimeters in diameter). It looks a bit like a flattened beer can, standing approximately 3 centimeters tall, and is full of electronic hardware.

The Khepera can be equipped with a host of sensors and actuators, though for our obstacle-avoidance experiment we need only one type of sensor and one type of actuator. The robot's eyes are eight infrared sensors positioned around its body; these are much simpler than our own eyes and serve to convey one basic piece of information: whether there is an obstacle close by (for this reason they're often called proximity sensors). The robot's actuators are two wheels attached to its bottom, allowing it to roam about. The Khepera thus sees the world through a ring of eight eyes, which provide it with a picture of the nearby surroundings. It is shortsighted, with the proximity sensors awakening only when the robot approaches an object, though it has the advantage of possessing a 360° view of the world. Being able to see its surroundings, the Khepera can *potentially* wander about by controlling its two wheels so that it doesn't crash into obstructions. I say "potentially" because there's still something missing: We need to program the robot's "brain"—the internal computer chip that guides its behavior—so that it correctly interprets the world it senses (the sensor readings), thereby commanding its wheels to change course when an obstacle looms ahead. And how do we program the robot's brain? We don't—we'll let it *evolve*.

We've already seen how we can evolve solutions to problems such as bridge building and analog-circuit design. Now we'll apply this process to evolve a small robotic brain for our Khepera, so that it can wander about without bumping into objects. Such an evolutionary experiment was carried out by Francesco Mondada, one of the Khepera's original designers, and his colleague Dario Floreano. So how do we go about evolving robotic behavior? Remember that the first thing to do when applying artificial evolution is to define exactly what you want to find, delineating the goal of the evolutionary process. That's pretty clear: Evolve a robot that exhibits obstacle-avoidance behavior. Next, you must design the artificial genome—the recipe that defines an organism in our evolving population (with each such organism being a candidate solution to the obstacle-avoidance problem). Since we're evolving small brains, why not use artificial neural networks? Remember from Chapter 3 that these are loosely based on the workings of real brains and are thus just the thing we need here. A simple artificial neural network that acts as the Khepera's brain might consist of just two layers: an input layer with eight neurons—one per proximity sensor—and an output layer with two neurons—one per wheel. The network receives the readings of the proximity sensors every one-tenth of a second, which it proceeds to process (as we saw in Chapter 3), outputting two motor commands to the wheels through the output neurons.

Having decided that a robot's genome is made up of an artificial neural network, what's next on the evolutionary agenda? We need to indicate the criterion by which the organisms will be judged, namely, their fitness. What makes for a good robot in our case? A good fitness definition is as follows: It wanders about without bumping into things. All we have to do is translate it into more precise terms: "Wanders about" is essentially the robot's speed, and "without bumping into things" can be measured through the proximity sensors. A better robot—with higher fitness—is one whose speed is higher *and* whose number of crashes is lower. Just how tricky defining fitness can be at times was evidenced in an early experiment where Floreano and Mondada measured speed by considering how fast the wheels rotated.

Evolution then found the following solution: a robot that spins about itself, like an ice-skater. This was actually a perfect solution with respect to the fitness definition in question: The wheels turn at a maximum rate, and, of course, the robot does not bump into any object! To correct this spinning problem, fitness had to be redefined to include not only the rate at which the wheels spin but also how far the robot advances. This example is quite instructive since it echoes a situation that we see in Nature: Evolutionary changes often provide totally unexpected solutions, such as fish that breathe air or birds that can't fly.

Now that we've nailed down the genome and fitness definitions, the computer can take it from here: It first creates a population of hundreds of completely random genomes—each representing a possible (neural network) solution to the obstacle-avoidance problem. Each of these random individuals, collectively comprising the first generation of robots, takes the "survival" test, that is, each is assigned a score representing to what degree it satisfies the fitness criterion. Now comes the selection-crossover-mutation phase: The computer selects individual robots for reproduction in accordance with their fitness, combines the lucky parents' genomes, and finally sprinkles some mutations here and there. Behold the next generation of robots. Next, it's simply a matter of iterating the above process: evaluating each individual's fitness score, selecting the winning couples, combining their genomes, and mutating.

You're probably imagining hundreds of robots running around, cheerfully bumping into things, some of which are bad (a chair) and some of which are good (a mate). Reality, however, is somewhat more mundane: There's actually only one robot. Maintaining a population of hundreds of robots would require a field the size of Yankee Stadium, with a price tag running into several million dollars. To cut down on costs, the population exists only virtually in a computer that is connected to but a single robot. To evaluate an individual genome, the computer running the evolutionary process places the neural network in question into the robot's brain—the computer chip that controls its behavior. Doing so completely erases the previous network that was there, and since the neural network is all that defines the robot's behavior in our case, we're now in possession of an entirely different individual. We then let this robot run around for a few seconds, evaluating its behavior so that a fitness score can be assigned. All other evolutionary operations—selection (reproduction), crossover, and mutation—are carried out by the connected computer; this is actually quite easy since it involves only the genomes, which are stored within the computer's memory in any case. The crux of the matter is that the evaluation phase is carried out in the *real world*. The robot roams about in a real-life environment, bumping into actual objects. (One of the advantages of the Khepera robot is its small mousy size, enabling researchers to perform their experiments on their desktops; larger cat-sized or dog-sized robots require an entire office to be sacrificed for the purpose of serving as training grounds.)

The setup we've just described does indeed yield robots that are able to carry themselves with grace, wandering about while shunning obstacles. It might take a while, though: If each robot is evaluated for 20 seconds, then evaluating a generation of 100 robots takes about half an hour. Since evolution might require a few hundred generations in order to produce good robots, the whole shebang might take a couple of weeks. But then again—who said learning to walk is easy? As we've seen, being able to move about is the first step that Dean the housemaid must learn to take. It is quite fascinating to see that one can use Darwinian evolution to produce neural networks—small artificial brains—that guide our robot.

Adaptive robotics is an active field of research these days; the underlying goal is to render robots more adaptive, able to handle dynamic situations and real-world environments. Researchers are working on training their robots to perform a host of tasks. For example, a group at the Massachusetts Institute of Technology looked into building a dog-sized robot with a mechanical gripper, which would stroll in and out of offices during the night; as it did so, the robot would collect empty beverage cans strewn about (or at least would try to) to be placed in some central repository. This machine might be considered one of Dean's ancestors.

Christof Teuscher, Eduardo Sanchez, and I developed a simple electronic robotic "brain" and used it in conjunction with Lego parts to build the Romero twins shown in Figure 4-2. Romero (Spanish for "pilgrim") is a cheap, no-frills

Figure 4-2   The Romero twins
COURTESY OF ANDRÉ ("CHICO") BADERTSCHER

robot: It has two wheels, two sonar sensors (the bulging "eyes" on the top, one of which is a sonar transmitter; the other, a receiver), and two visible-light sensors attached to the bottom panel and facing the ground. We used Romero in a simple experiment: evolving the robot to follow a broken trail using the bottom-panel light sensors. Sounds simple? Not with our cheap pilgrim: Romero cost us about $150 in all, much cheaper than most other robots (a Khepera costs a few thousand dollars). This inexpensiveness, however, comes with a price: Romero's sensors (sonar, light) and actuators (wheels, motors) are of relatively low quality. Thus, such prima facie simple tasks as following a trail or even just moving in a straight line are not trivial. You'd think it's just a matter of supplying the motors on each side with the same power, thus having the wheels turn at the same speed. Mechanical inaccuracies of the motors and wheels, however, will require a slightly different power to each motor, with constant power corrections due to friction, inertial forces, unevenness in the ground, slipperiness, and so on. (Walking a straight line is not trivial even for us: Consider how the police use this behavior as a test for inebriation.) More expensive robots have higher-quality sensors and actuators and are thus more accurate and more reliable.

The advantage of inexpensiveness is that you can have a real population of robots. While we only have twins for the moment, Richard Watson, Sevan Ficici, and Jordan Pollack from Brandeis University have been more (re)productive: They have an octet. They placed their wheeled creatures in a 1.3- by 2.0-meter pen with a lamp located in the middle, intending for them to perform phototaxis, that is, reaching the light from any starting point in the pen. They let their robots roam about and evolve autonomously in what the Brandeis researchers called *embodied evolution*, meaning evolution that occurs in a population of real robots. Because the pen contained a multitude of robots, there was an added difficulty, robot-to-robot interference (bumping into one another, blocking the light), which the robots had to overcome. And the outcome? Though the robots could not find the light at the end of the tunnel, they did evolve successfully to find the light in the middle of the pen.

Over the past couple of years researchers have become even more ambitious, with several robotics conferences now holding soccer tournaments—known as RoboCup—between teams of robots. The obstacle-avoidance setup contained only one robot, while the Brandeis octet simply had to avoid treading on one another's toes (or wheels). But when the robots in the population are to *cooperate*, exhibiting so-called *collective behavior*, the task becomes much harder. The soccer matches are quite amusing to watch; yet they also serve to remind us of the crudeness of our current-day robots. There is still much research that needs to be carried out. A robot team usually consists of three players, the object being to score as many goals as possible with a small puck. In many cases the robots act very strangely, exhibiting hardly any team play at all, and, at times, manifesting no discernible desire to score goals. The robot designers have noted that their teams often play much better at home than during the match at the conference site. While at their home institute, the robots become adapted to the quirks of their training field—its colors, landmarks, and what not. These small idiosyncrasies are used by their "brains" to improve their performance and score more goals. Alas, when they find themselves on a different field, they get "confused"; things are not exactly as they were at home. This just goes to show how hard it is at times to adapt (even for us, new situations may pose problems). Still, these robot competitions are highly entertainings, and winners do emerge. (While robot soccer matches may seem like a novel and perhaps strange idea, researchers have become quite used to them. So much so, in fact, that when my colleague Forrest Bennett, now with the FX Palo Alto Laboratory in California, tried to organize a soccer match among the participants of a recent conference on evolutionary computation, he had to reiterate repeatedly that he meant to hold a match between *humans*.)

The RoboCup tournaments are not held for mere ludic purposes. To quote from the RoboCup Web site (www. robocup.org): "The Robot World Cup Initiative (RoboCup) is an attempt to foster AI [artificial intelligence] and intelligent robotics research by providing a standard problem where a wide range of technologies can be integrated and examined. For this purpose, RoboCup chose to use the soccer game, and organize *RoboCup: The Robot World Cup Soccer Games and Conferences*. In order for a robot team to actually perform a soccer game, various technologies must be incorporated including: design principles of autonomous agents, multi-agent collaboration, strategy acquisition, real-time reasoning, robotics, and sensor-fusion. RoboCup is a task for a team of multiple fast-moving robots under a dynamic environment. RoboCup also offers a software platform for research on the software aspects of RoboCup." It is a serious game indeed.

One of RoboCup's main missions is to deepen our understanding of collective behavior, or, in other words, of how societies of robots function. This has many techno-logical implications. For example, there has been talk of sending hordes of robots into space to perform planetary exploration. This is a relatively cheap way of scouting a planet before sending in the human cavalry.

Researchers in adaptive robotics do not only have techno-logical objectives—such as building autonomous housemaids or space explorers—in mind. They also consider their artificial creatures as tools for better understanding biological organisms. We can learn much about natural societies by studying robotic ones. Insect behavior is another field of natural study that could benefit greatly from robotic experiments.

Barbara Webb from the University of Nottingham studied the behavior of female crickets locating males from the sound of their chirping. What kinds of paths do female crickets take when advancing toward the male (straight, zigzag)? How does the female behave when there are two male speakers? How can we explain the selectivity of females for certain male songs? "To investigate these issues," wrote Webb, "I designed and built a cricket robot based on relatively simple hardware, principles and algorithms. The behavior of this robot, I hoped, would have much of the complexity of the insect itself when confronted with a similar environment. I also hoped to cast some light on the neurobiology that may underlie cricket behavior."

Webb built a "female" cricket robot that was allowed to move in a small arena toward a speaker emitting a noise modeled on the sound produced by a male cricket. The path "she" took to the speaker zigzagged much like the route of a real female cricket moving toward a mate. Webb then added a second speaker, another male voice; "To my surprise," she wrote, "the robot seemed to have no problem making up its mind (so to speak) and went almost directly to one speaker or the other." Like a real female cricket, the robot chose one calling male from among several. Webb concluded carefully that "These results as a whole are encouraging, although of course the robot's success does not in itself prove that real crickets work this way. Nevertheless, it does suggest some alternative interpretations of neurophysiological and behavioral results." Our robots still have a long way to go until they match the complexity of their living inspirations. But already they serve to advance the march of science.

Small, versatile insect robots might perform many useful tasks, one example of which is search and rescue missions. Imagine sending a horde of such "robosects" into the ruins of a building destroyed by an earthquake. Their minute size will allow them to scour nooks and crannies, and—being equipped with delicate sensory apparatus—they will be able to locate survivors and communicate their positions to the rescue team outside.

As I remarked earlier, insects are quite intelligent little critters. As indeed are all creatures great and small. We can learn *from* them how better to build robots, which in turn can help us learn *about* them.

Now, Dean, I'd like to explain to you something about Rusty the dog....

CHAPTER 5

It's a Bit Warm in Here

You arrive at a friend's house for dinner on a cold winter night. He introduces you to the several guests already present, one of whom has an eerie look about him, which you cannot quite pinpoint. The heating is obviously on, it's freezing outside, and a few minutes into dinner that weird-looking fellow interjects a remark that almost makes you choke: "The temperature in this room is 3.2° above the acceptable level." Huh?

Your host, a cutting-edge robotics researcher, intervenes at this point. "Got you fooled there for a moment, didn't I?" he announces with pride. "Andy here is in fact a humanoid robot, the latest result of my research efforts. Its body is a technological marvel—almost indistinguishable from a real human body, though as you can see, once it opens its mouth, well, it's a dead giveaway: The illusion of humanity immediately breaks off. We still have much to do down at the lab...."

Why is Andy's remark so out of place? Because it's too mechanical. If we humans feel that the temperature is "above our acceptable level," then we just come out and say something like: "It's a bit warm in here." The many tongues "spoken" by computers are all characterized by being precise and explicit, whereas humans tend to be imprecise and ambiguous—though we have no problem in handling this vagueness. The expression "it's a bit warm in here" will be perfectly understood by your host, even though it includes such imprecise terms as "a bit" and "warm." The response will probably be vague as well—something like "I'll turn down the heating a little bit"—which will be quite all right with you. Our brains constantly handle such imprecision with no discernible effort.

Computers are crisp creatures, whereas we humans are fuzzy. The question that arises is: Can we narrow this gap? There are two possible ways to do so. The first is by forcing humans to behave more crisply, to be precise and unambiguous; this is exactly the stance that computer programmers must assume. Since they converse in the computer's tongue, they must be very "mechanical," avoiding the use of imprecise concepts (if not, the computer will "complain" by issuing an error message). There is another way to narrow the human-computer gap, though, which is much less "painful" to us: having the computer behave in a fuzzier manner. Can this be done—can we "fuzzify" computers? Yes, by using *fuzzy logic*.

Fuzzy logic made its appearance in 1965, when Lotfi Zadeh, then at the University of California, Berkeley, published a paper entitled "Fuzzy Sets." The field has gained prominence over the past two decades, finding its way beyond academic circles into industry. Many successful commercial applications—fuzzy systems—have been built to date.

To understand the idea behind fuzzy logic, let's first talk about sets. A *set*, or *class*, is a collection of elements that belong to a *universe of discourse*—this latter is a mathematical term that refers to the entire ensemble of objects that make up the "world" under consideration. For example, let's suppose that our universe of discourse is the collection (set) of all cartoon characters. An element in this set is an individual character, such as Bugs Bunny or Popeye. However, we need not necessarily consider either the entire universe or an individual member. We can also speak of subsets, that is, sets that include only part of our world. Some possible subsets are animals (including, among others, Bugs Bunny, Daffy Duck, and the Road Runner), superhumans (such as Superman, Spiderman, and Wonder Woman), humans (Lois Lane, Bart Simpson), short characters (Yosemite Sam, Tweety the bird, and Jerry the mouse), and tall characters (Superman, Batman).

When talking about classical logic—the kind usually employed by computers—then the sets in question are crisp. First and foremost this means that an element either belongs to the subset in question—or not. Superman belongs to the set of superhumans, whereas Bart Simpson does not. We can think of each set as having an appointed *membership daemon*—a cheerful little fellow who is an expert on his set. Present him with any individual from the universe of discourse (cartoon characters, in our case), and he'll immediately tell you whether that individual is in or out. The daemon has this irksome quirk of uttering only numbers. For example, when you ask the membership daemon of the animals set: "Is Superman a member?" then his answer is: "0." Persisting, you ask him: "Is Daffy in?" and he answers: "100." He uses a simple code: Instead of replying with "yes" or "no," like a well-mannered daemon he prefers to give you the membership percentage—100 percent for a member and 0 percent for a nonmember. With crisp sets, these are the only two possible answers: It's either in all the way, or totally out—nothing in between.

There are many things you can do with crisp sets, which is quite fortunate. They enabled us to build the powerful computers we see around us today. Without running the whole gamut of possibilities, let's just examine three basic crisp-set operations, known as *union* (also called *or*), *intersection* (also called *and*), and *complement* (also called *not*). Union amounts to constructing a set whose members come from *either* of two different sets. The union "animal characters *or* short guys" is itself a set; you can think of it as a club where entry is gained by obtaining a yes (100 percent) answer from *either* the animals daemon *or* the short-guys daemon (or both). Tweety (animal) and Bart Simpson (short) are in; Superman is out (neither animal nor short). Intersection is also a club where entry is gained by questioning two daemons. However, this time *both* the first *and* the second must grant their permission, that is, say yes. The intersection set "animal characters *and* short guys" would thus include Tweety (animal *and* short) but not Bart Simpson (short, but not animal). Finally, the complement set contains all elements that do *not* belong to a given set, that is, for which the daemon answered "no." The animals complement is thus a set that contains all nonanimals: Superman, Bart Simpson, Yosemite Sam, and the rest of the gang. Note that we always remain within the universe of discourse, that is, the "world" in question: The animals complement set does not include *every* imaginable object in the world at large that isn't a cartoon animal; it contains only every *cartoon* character that's not an animal.

There are two fundamental principles, or laws, of crisp sets that date back to Aristotle: the law of contradiction and the law of excluded middle. The first refers to the fact that if you consider a set plus its complement, then you end up with the entire universe of discourse: Human cartoon characters plus nonhuman cartoon characters include all cartoon characters, which is the world in question. The law of excluded middle means that an element is either in a set or in its complement—but not in both. Bugs Bunny is either in the animals set or in the nonanimals set, but not in both sets simultaneously. These two laws seem quite obvious in view of what we've presented about crisp sets: You're either a member or not—nothing in between. Now it's time to make things fuzzier.

With fuzzy sets our membership daemons become much more talkative: Although they're still interested only in numbers, they can now respond with anything between 0 percent and 100 percent (and not just the two extremes). This is in fact the crux of the matter, embodying the manner in which we humans tend to think. Take, for example, the short-guys set: Jerry the mouse is definitely 100 percent in, but what about Bart Simpson? He'd be better described (in our terms) as *shortish* rather than as *short*. With crisp sets we cannot accommodate this distinction; you're either in or out, buster; make up your mind. But the fuzzy membership daemon has no problem when you ask him about Bart's membership status in the short-guys club: "75 percent," he answers. A fuzzy membership value measures the extent to which an element is a member of the (fuzzy) set, or, in other words, how similar the object is to the imprecise property in question. Bart Simpson, while not 100 percent short, is still very much so— to the extent of 75 percent.

With fuzzy sets we must define de novo the meanings of operations such as union, intersection, and complement. Things are a bit more complicated here since we cannot form a view of a fuzzy set as we can with a crisp one. It's easy to picture the crisp set of short cartoon characters by simply listing all its members (Bart Simpson, Tweety, and so on). What does the fuzzy, short-guys set look like though? Does it contain a version of Bart Simpson reduced to 75 percent of his original size, or perhaps, more gruesomely, Bart minus some body parts? Actually, we *can't* form a nice crisp picture with fuzzy sets, but we *can* account for the workings of the fuzzy membership daemons. Remember that the union of two sets—whether crisp or fuzzy—is itself a set. The union of two fuzzy sets is thus a fuzzy set in its own right, and, as we've just seen, we cannot create a picture of it. But, since this union *is* a set, we *can* talk about its membership daemon. The fuzzy union "animal characters *or* short guys" is itself a fuzzy set, and as such will have its own, third daemon. Now, if I come along and ask this daemon, "What's Bart's membership status; that is, to what degree is he a member of this union?" what will the daemon answer? Well, this union daemon will ask both the animal-characters daemon and the short-guys daemon for Bart's membership status, and then he gives as a final answer the *higher* of the two responses. Since the animals daemon says "0 percent" (despite his behavior, Bart's no animal) and the short-guys daemon answers "75 percent" (Bart's shortish), the union daemon's job is quite simple. His final word is "75 percent"—the higher of the two responses. Notice that there's no underlying image of the union set with floating Bart parts: It's all about membership daemons. There may not be a nice picture of the union fuzzy set, but that's no problem at all. The union daemon can still answer *any* question about membership. The intersection set also gets assigned a daemon who performs the same interrogation as his union colleague but returns the *lower* of the two as an answer. The "animal characters *and* short guys" daemon, when asked about Bugs Bunny, will obtain the answer "100 percent" from the animals daemon, and, say, "65 percent" from the short-guys daemon; his final say will then be "65 percent" (the lower of the two answers).

Why did we combine the daemons' answers in this particular way? Actually, there are other possibilities for doing fuzzy union and intersection. The case for fuzzy sets is not as straightforward as it is for crisp sets. The choices we presented are those most commonly used in the field today. To understand why this makes sense, think of doctors and golf players and assume that membership in each group is fuzzy (part-time membership can thus be taken into account). Suppose they unite, that is, form a union (in the fuzzy-set sense). It seems reasonable to compute each person's membership value by considering the higher of the two original values. If you're a 40 percent doctor and a 70 percent golf player, then you're a 70 percent member of the new union of the two. (Adding the two, for example, won't do. You end up with 110 percent membership!)

The fuzzy membership value for the complement set is obtained by subtracting the daemon's answer from 100: Bart Simpson is a 75 percent member of the short-guys set, which means that he's a 25 percent member of the non-short-guys set (the tall guys). This shows an important difference from crisp sets: An element can belong to both a set and its complement. Bart is mainly shortish, but he is also somewhat tallish. Thus, we have broken the law of excluded middle. In fact, the law of contradiction is also broken: The union of a set and its complement does not make up for the entire universe of discourse. Since Bart is a 75 percent member of the short-guys set and a 25 percent member of the complement tall-guys set, he's a 75 percent member of the union of the two (remember, with union we take the higher of the two values); this does not amount to the universe of discourse, which contains one bona fide, fully 100 percent Bart.

This last part may sound a bit confusing, but then so is our own language. When we say that somebody is tall, what exactly do we mean? That he's six feet tall? Five feet and eight inches? Six feet and three inches? The point is we usually don't really care about the precise value (unless we're talking about basketball players). If we wish to refine our statement, we usually resort to the use of linguistic descriptions such as "rather tall," "quite tall," "tallish," or "very tall." The great thing about fuzzy logic is that it can handle such expressions by using so-called linguistic variables. This means that one can use words rather than numbers. Indeed, we have already encountered such linguistic tags; they're the names of the sets we've been using (animal, human, short, tall).

To take a different example of a linguistic variable, consider the temperature in your room. The "variable" part means that the temperature can vary from one moment to the next. The "linguistic" part means that the value at any given moment is not a number (23° Celsius) but rather a word, say, one of the following five possibilities: cold, cool, OK, warm, hot. Fuzzy logic enables the computer to handle such imprecise, human terms by employing membership daemons like the ones we describe above.

When building a fuzzy system, a major job of the system designer is to imbue the daemons with their knowledge, that is, she must decide how each daemon will compute its membership values. In the temperature example there are five dae-mons—cold, cool, OK, warm, hot—and each one will react differently to a given temperature. Suppose that your thermometer shows 18° Celsius: Is that OK? Cool? Perhaps warm? Well, we're not talking about crisp logic here; it's actually a bit of all three. The "cool" daemon might consider the membership value of 18° to be 50 percent, the "OK" daemon might consider this temperature to have a membership value of 75 percent, and the "warm" daemon might say 30 percent. Both the "cold" and "hot" daemons consider 18° to be completely out of their depth: They both answer 0 percent if asked to what extent 18° belongs to their set. You could think of this as meaning that the temperature is mainly OK, but it's also somewhat cool, and—to a lesser extent—a bit warm.

If the computer can't even make up its own mind as to whether it's currently cold, OK, or warm, then what good is all this? First, let's not forget that this five-daemon, fuzzy setup has not resulted in complete indecision. We're simply faced with three differing opinions, with a penchant toward "OK."

Now it's time to talk about how fuzzy logic is used to build *fuzzy systems*. I'll concentrate on one important domain known as *control systems* in which fuzzy logic has been exhibiting its most spectacular success. A control system is one that automatically (without any, or hardly any, human intervention) controls another system—usually mechanical or electrical (the latter is the *controlled* system).

Since we've been talking about room temperature, let's see what it takes to build a fuzzy air conditioner. This machine has a small computer within it—the control system—that controls the operation of an internal motor—the controlled system. Once you set the desired room temperature, the air conditioner must take care to maintain it. How can it do so automatically? First, the system must have some means of sensing the environmental temperature, which is quite easy to do. We just plug in a thermometer, which supplies the control system with the *input*—the current temperature. Next, the fuzzy control system decides whether any action must be taken: Is it too cold or too hot? If either of these holds true, then the control system sends an *output* message to the internal motor, directing it to increase its speed (room is too hot) or decrease its speed (room is too cold).

We're now faced with somewhat of a problem. The control system is fuzzy, yet the input and output are crisp. The input is a precise temperature reading, and the output must control a mechanical motor, which doesn't have a single fuzzy bone in its body. All it wants to know is the rpm value (revolutions per minute or how many times must it turn each minute; this is like the rpm indicator that you find in many cars). We obviously don't want to get rid of our fuzzy control system. Isn't that the whole point? After all, we have been insisting on how fuzzy systems are better suited for human use. So what can we do? How can we "stitch" the crisp-fuzzy borders? The solution is to pass the input through a "fuzzifier" and pass the output through a "defuzzifier." The fuzzifier transforms crisp numbers into fuzzy sets; we've already seen how it works when we handed our five daemons (cold, cool, OK, warm, hot) the temperature reading (18°), allowing each of them to come up with a membership value for it. The defuzzifier transforms a fuzzy output into a crisp number. The air conditioner now works as follows: Once a minute a temperature reading is taken, which is fuzzified by the fuzzifier and handed to the fuzzy control system. The fuzzy control system then makes a decision and sends out a fuzzy output. The output is transformed by the defuzzifier into a crisp rpm value and sent to the motor, which corrects its spinning in accordance with the command it has just been given. The cycle thus contains five phases: temperature reading → fuzzifier → fuzzy control system → defuzzifier → motor control.

At the heart of the fuzzy control system lies an *inference engine*, which makes inferential decisions in accordance with a set of *fuzzy rules*. These rules are of the form *if such and such, then do this and that*. Our air conditioner control system might contain the following five rules:

1. If the room is hot, then substantially increase motor speed.
2. If the room is warm, then slightly increase motor speed.
3. If the room is OK, then do nothing.
4. If the room is cool, then slightly decrease motor speed.
5. If the room is cold, then stop motor altogether.

Note that not only is the first part of a rule (known as the antecedent) fuzzy, but so is the second part (known as the consequent). The inference engine uses these five fuzzy rules to arrive at a fuzzy decision.

When a temperature reading of 18° is taken, it is first fuzzified by having each of the five daemons decide on the membership value for the set under its responsibility. We have already taken note of their answers: cold, 0 percent; cool, 50 percent; OK, 75 percent; warm, 30 percent; and hot, 0 percent (that is, 18° is not a member of the cold set, is a 50 percent member of the cool set, and so on). This fuzzy input is handed to our five-rule fuzzy control system, causing the *activation* of three rules—the "cool," "OK," and "warm" rules. A rule is activated when its antecedent is "awakened," that is, the condition specified in the first part is fulfilled to some extent (above 0 percent). In this case the rule is said to "fire," meaning that it "shoots" its second part—the consequent. Obviously, we cannot send three different commands to the motor; in fact, each of the three rules has given but a *recommendation*. These recommendations do not carry equal weight: the "OK" rule has been activated much more strongly than the "warm" rule since its antecedent condition (the "if" part) is fulfilled to a greater extent (it's "OK" to the extent of 75 percent but "warm" only to the extent of 30 percent).

Stronger rules pack more punch, meaning that their consequents (the "then" parts) are ascribed larger weights when considered by the defuzzifier. This then combines the three actions recommended by the three rules into one single action: a crisp control command to be sent to the motor. When combining the three recommended actions, the defuzzifier places the most emphasis on the "OK" rule's recommendation ("do nothing"), less so on the "cool" rule's recommendation ("slightly decrease motor speed"), and the least emphasis on the "warm" rule's recommendation ("slightly increase motor speed"). All activated rules have their say in proportion to their strength (similar to the U.S. House of Representatives where more populous states have more representatives). The bottom line in this case will most likely be to slightly reduce the motor speed, thus causing a small rise in room temperature (because you've probably set your desired temperature to about 19° Celsius).

This is quite a simple fuzzy control system. A more sophisticated air conditioner might take into account the readings of three thermometers placed in three different locations in the room. This would allow the system to maintain finer control of the temperature. It would entail, though, a more sophisticated fuzzy system, with more complex rules, such as "If thermometer No. 1 reads warm *and* thermometer No. 2 reads hot *and* thermometer No. 3 reads OK, then increase motor speed by an intermediate amount."

Designing a fuzzy system is no mean feat. There are many issues that come into play, such as: What are the relevant linguistic variables? How do the membership daemons compute their membership values? What rules should go into the fuzzy system? How should they be combined? In many cases the system designer is either an expert in the field or works in close collaboration with one. For example, an air-conditioning expert is needed in order to design the fuzzy control system discussed previously.

Despite these difficulties, many successful commercial applications of fuzzy systems have been developed over the past two decades. You'll find such systems in cars, television sets, washing machines, chemical plants, and medical systems, to mention but a few. One of the most celebrated applications is the Japanese Sendai Subway Automatic Operations Controller. Introduced in 1986 by Hitachi, it implements the strategies of experienced operators in the form of fuzzy rules; the resulting automatic subway is faster (since it better controlled the braking apparatus), smoother, and more energy efficient than human-driven subways. Home appliances (especially Japanese ones) have also seen a proliferation of fuzzy logic. For example, one video camera has an auto-focus fuzzy system in which the input is an approximate measure of sharpness, and the output is a motor-speed control. This system has a total of thirteen rules, such as: "If the sharpness is high *and* its differential is low, then the focus motor speed is low." This fuzzy control system improves the focusing quality and reduces focusing time.

More recently, system designers have been seeking to avoid the manual design of fuzzy systems by using some of the methodologies described in this book, such as artificial neural networks and evolutionary computation. One major field of application for such evolutionary-fuzzy and neural-fuzzy systems is medicine.

Medical professionals are confronted daily with numerous problems from diverse walks of (medical) life. They all exhibit an underlying commonality: searching for a good solution among a (usually huge) number of possible solutions. Whether trying to pry out signs of malignant cancer in cell biopsies or looking for irregularities in EEG signals, the basic problem is that of sifting through a welter of candidate solutions to find the best possible solution. As in any other area of modern life, computers are omnipresent in medicine, from the hospital accounting computer to the high-end MRI scanner. In particular, computers are used as tools to abet medical professionals in resolving search problems such as those involved, for example, in medical diagnosis. Diagnosis is the process of gathering information concerning a patient and interpreting it according to previous knowledge, as evidence for or against the presence or absence of disorders. (This should be distinguished from prognosis, where the objective is to predict the future development of the patient's condition.)

Take breast cancer, for example, where the treating physician is interested in diagnosing whether the patient under examination exhibits the symptoms of a benign case, or whether her case is a malignant one. My colleague Carlos-Andrés Peña-Reyes and I evolved fuzzy systems for diagnosing breast cancer.

A good computerized diagnostic tool should possess two characteristics, which are often in conflict. First, we want it to attain the highest possible *performance*, diagnosing correctly as many of the presented cases as possible as being either *benign* or *malignant*. Moreover, we'd like to have the system output a so-called degree of confidence: not merely a dichotomous benign-or-malignant diagnosis, but also a numeric value that represents the degree to which the system is confident about its response. Second, we want the diagnostic system to be human-friendly, exhibiting so-called interpretability. This means that the physician is not faced with a black box that simply spouts answers (albeit correct) with no explanation; rather, the system should provide some insight as to *how* it derives its outputs. Fuzzy systems fit both these bills: They can attain high performance, and the fuzzy rules can be understood by doctors. In contrast, artificial neural networks operate in a "black-box" manner, with their knowledge embodied mostly within the synaptic weights; these weights are utterly incomprehensible to a human being. Although methods exist for "extracting" high-level rules from artificial neural networks, fuzzy logic is often a better choice where interpretability is of the essence.

Peña-Reyes and I looked into the so-called Wisconsin breast cancer diagnosis problem. Breast cancer is the second most common cancer among women (after skin cancer), and its early diagnosis is thus of primordial importance. One common outpatient procedure is known as fine needle aspiration (FNA), which involves using a small-gauge needle to extract fluid from a potentially cancerous breast mass. It is a cost-effective, nontraumatic, and mostly noninvasive diagnostic test that obtains information needed to evaluate malignancy.

Interested in diagnosing breast cancer based solely on an FNA test, experts at the University of Wisconsin Hospital identified nine visually assessed characteristics of an FNA sample considered relevant for diagnosis:

1. Clump thickness
2. Uniformity of cell size
3. Uniformity of cell shape
4. Marginal adhesion
5. Single epithelial cell size
6. Bare nuclei
7. Bland chromatin
8. Normal nucleoli
9. Mitosis

If you're not a doctor, then the list above is probably incomprehensible, as it was for Peña-Reyes and me (we're *not* medical doctors). Fortunately, this is no obstacle at all (we'll see why in a moment). Once a sample is taken, specialists assign a number between 1 and 10 to each of these characteristics. The University of Wisconsin experts created a database of 683 cases with known diagnoses. This database is a list containing 683 rows, with each row corresponding to the nine measured items pertaining to one sample (from one patient). Each row is accompanied by a diagnosis provided by specialists: benign or malignant.

	ITEM					
CASE	(1)	(2)	(3)	...	(9)	DIAGNOSTIC
1	5	1	1	...	1	benign
2	5	4	4	...	1	benign
.	.	.	.	...	.	.
.	.	.	.	...	.	.
.	.	.	.	...	.	.
683	4	8	8	...	1	malignant

The first case, for example, stipulates that when item number 1 (clump thickness) has a value of 5, item number 2 (uniformity of cell size) has a value of 1, item number 3 (uniformity of cell shape) has a value of 1, and so on until item number 9 (mitosis) which has a value of 1; then the case is benign.

Once the database of 683 cases was completed, the problem was transferred from the domain of medicine to that of computing. This is why computing scientists such as Peña-Reyes and I could tackle this medical-diagnosis problem: It's all numbers. In all honesty, we have never performed fine-needle aspiration, nor even attended one. This situation is typical of many problems in multitudinous domains: Once the relevant data (be they medical, aeronautical, or whatnot) have been transformed into a computerized form, the problem becomes a computational one. This is quite fortunate: It allows computing professionals to help out in domains in which they have little or no knowledge.

Having obtained the Wisconsin database on the Internet (the list of 683 cases), Peña-Reyes and I now faced the following computational problem: finding a small set of fuzzy rules that could account for as many of the above cases as possible. We could not ascertain in advance how hard the problem was; it might have turned out, in fact, to be trivial. Clump thickness alone, for example, might have turned out to be a sufficient indicator of benignity or malignancy. That would mean that there would have been a very simple (fuzzy) rule: "If clump thickness is low, then the case is benign" (where "low" is defined in a fuzzy manner, that is, it is a fuzzy set with a membership daemon). In the real world, of course, things are not that simple (including this problem).

Constructing good fuzzy systems by hand is hard, so Peña-Reyes and I evolved them. Our evolutionary-computation scenario contained a population of individuals, each one of which was *an entire fuzzy system*. We fixed in advance the number of rules these fuzzy systems included. There is a tradeoff here: A fuzzy system with fewer rules is easier to grasp, but it often performs less well; a fuzzy system with more rules is usually harder to understand, but it usually performs better. By "performance" I mean the number of cases of the total 683 that the fuzzy system classifies correctly (in agreement with the database) as benign or malignant.

Each individual in our population was a complete fuzzy system, whose number of rules we fixed at the outset. (We ran the evolutionary simulation numerous times, going from one-rule systems up to seven-rule systems.) Each fuzzy system in our evolving population was specified by a genome that defined two things: (1) the rules themselves (if we fixed the number of rules to be two, then the genome allowed for two different rules) and (2) the membership daemons for the nine items on the FNA list. Each item could be either "low" or "high," but the meanings of "low" and "high" changed from item to item: "Low" clump thickness is not the same as "low" marginal adhesion. But deciding how to define (fuzzily) these nine different forms of "low" was not for us: We left that, as well as finding the rules, to evolution.

The fitness of an individual in our population (a fuzzy system) was quite easy to define for this problem: It was simply the system's performance, namely, the number of cases (of the 683) it classified correctly as benign or malignant.

The evolutionary setup in place, we now sat back and let the simulations run on our computers. To our delight, we observed the emergence of highly successful systems, ones whose classification performance was as high as 98 percent. This means that of the 683 cases in the database, only about

15 were classified incorrectly (benign instead of malignant or vice versa). We found that even with only one fuzzy rule we could obtain a performance rate of 97 percent. Here's one of our evolved single-rule systems:

> If clump thickness is low *and* uniformity of cell size is low *and* bare nuclei is low *and* normal nucleoli is low, then the case is benign; otherwise the case is malignant.

Not only is this a high-performance system, it is also interpretable (at least if you're a doctor). We can thus combine evolutionary computation and fuzzy logic to solve important, real-world problems.

Encouraged by our success with evolution, Peña-Reyes and I have turned more recently to *coevolution*: the simultaneous evolution of two or more species with coupled fitness. On this subject Darwin wrote: "I can see no limit to the amount of change, to the beauty and infinite complexity of the coadaptations between all organic beings, one with another and with their physical conditions of life...." Evolution of organisms to their "physical conditions of life" has been our main source of inspiration for evolutionary computation, wherein the "physical conditions" within our simulated world are embodied as the fitness. Coevolutionary computation is inspired by the adaptation of organisms "one with another."

Coevolution in Nature comes in two flavors: competitive and cooperative. In competitive coevolution the fitness of an individual is based on direct competition with individuals of *other species*. Increased fitness of one of the species implies a diminution in the fitness of the others. An example is two-species predator-prey coevolution, where the evolution of better survival strategies of one species (in terms of speed, adroitness, stealth, etc.) implies an immediate survival challenge to the other. When a population of lions on the African savanna evolves faster legs, it changes the environment for the roaming zebras. To survive and reproduce, they must now evolve to run faster; this in turn means that the lions will have to run yet faster, thus creating what biologists call an "arms race," a term borrowed from a nasty form of human-technology coevolution. In cooperative coevolution, the individual species collaborate rather than compete, with an increase in the fitness of one species implying an increase (rather than a reduction) in the fitness of the other. Think of the cozy relationship that has evolved between honeybees and flowers. The flowers supply the bees with nourishment (nectar) while the bees abet the flowers' pollination (reproduction).

Both forms of coevolution in nature—competitive and cooperative—have inspired a number of coevolutionary algorithms over the past few years. Peña-Reyes and I have designed an evolutionary algorithm dubbed Fuzzy CoCo (for Fuzzy Cooperative Coevolution) that we applied with success to the Wisconsin breast cancer diagnosis problem. As we saw above, in the standard (single-species) evolutionary setup, an individual in the population makes up an entire fuzzy system, its genome clumping together both the fuzzy rules and the membership daemons. In Fuzzy CoCo, we "split" the genome in two, defining two species: one containing fuzzy rules and the other containing membership daemons. An individual within a species now can no longer solve the problem by itself, since it is no longer a complete fuzzy system. So how do we ascribe fitness to species members? By having them cooperate. In every generation, choice (fit) members of each species are selected to act as "cooperators"—representatives of their species—that are combined to form complete fuzzy systems, which can then be evaluated for fitness. Both sides reap the benefits of a good collaboration (or suffer the consequences of a bad one) in that the resulting fitness is ascribed to both individuals of the cooperating team. The upshot of our coevolutionary scenario was the evolution of yet better breast-cancer diagnosis systems—as high as 99 percent classification performance—which took less time to evolve (while gaining 1 or 2 percent may not seem impressive, remember that this implies several more correctly classified cases; at this performance level, every additional case is hard-won).

I'd like to end this chapter with a question that often arises among students encountering fuzzy systems for the first time. I believe this question might also serve to further defuzzify our understanding of the fuzzy idea: Isn't fuzziness the same thing as probability? What's the difference between saying that the room is within the "warm" set to the extent of 50 percent and saying that there's a 50 percent chance that the room is warm? I'll answer this by using a well-known example in the fuzzy literature. Suppose you're lost in the desert, and, having gone a few days without drinking, you come upon two bottles. The liquid contained in them is unknown to you, and in fact may not even be potable. The first bottle is labeled "membership value in set of potable liquids: 90 percent" and the second bottle is labeled "probability that liquid is potable: 90 percent." Which one would you choose to drink? A membership value of 90 percent means that the contents of the bottle are fairly similar to perfectly potable liquids (such as pure water)—perhaps with a bit of sand added. A probability, though, means something entirely different: There is a 9 in 10 chance that the contents of the bottle will turn out to be perfectly potable, and a 1 in 10 chance that the contents will be deadly. I'd say you're better off with the fuzzy bottle (a 1 in 10 chance of dying isn't very cheerful, after all).

There is another facet to this example, and it has to do with the idea of *observation*. Suppose you examine the contents of both bottles with a chemical toolkit and find out that the fuzzy bottle contains beer and the second bottle contains cyanide. *After* observation the potable-liquids membership daemon will still return 90 percent as a membership value when presented with the beer—no change there. Knowing that the second bottle contains cyanide, on the other hand, *has* changed the probability that the liquid is potable: from 90 to 0 percent. It's just like a horse race: Before the race a horse might have a 1 in 4 chance of winning, whereas afterward it's either 1 in 1 (if it won) or 0 (if it lost). Fuzziness and probability are thus two different things; the former is about similarities of objects to imprecisely defined properties, while the latter is about chances of events taking place. Clearly, a crisp distinction.

CHAPTER 6

Softening Hardware

In the early days of computing—back in the 1940s—computers were huge, room-size beasts, full of blinking lights and ambulant switches (you've probably seen the likes of them in old movies and TV programs). Your job as a programmer was to convince these mighty behemoths to do your bidding, that is, to make the computer solve a given problem. As the world was at war in those days, "your" bidding was more likely to be that of your commanding officer, and the task at hand could be anything from computing artillery firing tables to cracking the enemy's secret codes.

A computer programmer of the times combined some of the skills of a mathematician, a telephone switchboard operator, and a car mechanic. A program is basically a sequence of simple instructions that the computer can follow in a mechanical manner; it consists of very simple operations, such as multiply two numbers, add the result to a third number, store this last value in some corner of the computer's storage space (aka memory) so it can be retrieved and used at some later time, and so on.

First off, you put on the mathematician's hat and try to come up with the program: What operations will you use? In what order? What comprises the *input*, that is, what must you feed into the computer in order to facilitate its task (inputting the old enemy code might help the computer crack the new one)? And—let's not forget—what exactly do you want as *output*, that is, what is the response you're expecting? Having figured out the answers to all these questions, you're now in possession of a procedure for solving your problem: the program.

In today's computing world, being in possession of the program means that you, the programmer, can sit back and relax: The computer will pick it up from here. But let's not rush things; we're still in the 1940s. It's time to put on the switchboard operator's hat. What you have to do now is go over to your beloved leviathan and modify its entrails. You must connect wires between various components of this giant switchboard, and set several switches, so that you end up with a machine that will chug along with but a single purpose in life: to fulfill your deepest desire, which is, of course, to carry out your masterpiece of a program (in computing jargon this is known as "running" or "executing" a program, though the latter term would probably best be avoided so long as we're at war).

From time to time you must put on your car mechanic's hat (in all likelihood, more of a greasy cap) since the computer doesn't always make it all the way through the program. Nothing illustrates this better, in my mind, than the famous story of the first computer bug. In 1947, naval officer Grace Hopper (who ended her illustrious career holding the rank of rear admiral) was working with one of the few computers in existence at the time, the Mark II, which refused pertinaciously to run her program correctly. After much ado, which consisted of checking and rechecking every possible cause of failure, she finally started taking the machine apart, only to find a dead moth within its innards. The log book from those car-mechanic days contains an entry dated September 9, 1947, reading: "1545, Relay #70 Panel F, (moth) in relay"; next to the entry is the moth itself held in with wide Scotch tape. A few days later someone added: "First actual case of bug being found." Since that day computer programmers refer to errors in a program as "bugs" and to the tedious task of finding these errors as "debugging." Alas, though moths are rarely the cause these days, bugs are nonetheless omnipresent.

Programming one of those early-day computers meant that you actually had to *change the machine itself*—the hardware—so that it became adept at running your program. While not a simple matter by far, the ability to wire the machine enabled its *hardware* to *adapt* to your needs. In the 1950s the infant computer industry realized, however, that this approach was too cumbersome and error-prone. Modifying the machine's innards every time you wanted to run a program was a tedious business: enter the stored-program computer.

What is a stored-program computer? The idea seems quite simple, actually, in retrospect: The computing machine is divided into two parts—the *processor* and the *memory*. The processor is the component that actually processes—or runs— the program; it is the unit that ceaselessly performs that drudging task of running the instructions one after the other, just as ordained by you, the programmer. And from where does it fetch the program? You've got it: from memory. This is a physically separate element of the computer that is basically a large storehouse that acts as a repository for programs (yours, as well as those of many other programmers, for sharing was the rule back in those days).

Our computer now has two different, physically separate components: processor and memory. So what? Well, the good news is that we can get rid of both our telephone-switchboard and car-mechanic hats! Whereas before, running a program meant digging into the computer's entrails so that it essentially *became* your program, now all you have to do is place your program in the storehouse—the machine's memory. The computer operates by having the processor fetch the program from memory (perhaps not even in its entirety, but just the next few instructions needed), after which it proceeds to run it. Hail the age of general-purpose computers.

Why are such computers called *general purpose*? Different computers have different processors. Each processor is capable of executing a rather small number of instructions— add, subtract, store result, fetch result, and so on. This is known in computing parlance as the *instruction set.* The point is that while the number of basic instructions recognized by the processor is quite small, one can design *sequences of instructions*—perhaps very long ones—resulting in a rather complex program. This can be compared to Lego bricks. Though we may be in possession of but a small number of different kinds of bricks (the instruction set), we can none-theless erect a wide variety of beautiful edifices (the pro-grams). Nature also provides her own example. Even though there are only twenty amino acids—the basic building blocks of life—the number of possible sequences of such acids, known as proteins, is huge. We can thus supply the computer with any imaginable program (provided we use only "legal" instructions taken from the instruction set)—from gloomy account balancers to high-pressure games—by simply stor-ing the program in the machine's memory, whereupon the processor will pick it up and run it. This is why we call such computers "general purpose." They are no longer hardwired to behave in a specific manner, accomplishing but a single task. Rather, they are designed to behave in a very general manner, accepting a recipe (program) that they will then proceed to follow (run). You can think of this as the difference between an imaginary cooking machine and a cook: The cooking machine is "hardwired" to bake an apple pie, and you must completely rewire it if you wish it to bake chocolate chip cookies. The cook, on the other hand, is "general purpose": Give him any recipe, and he'll bake it.

The age of general-purpose computing, which came to be in the 1950s, is still upon us today. While processors got faster (in fact, a whole lot faster) and memories grew larger (indeed, much larger), it's still your basic stored-program, general-purpose computing. This new wind brought along with it what is the most basic dichotomy in computing—that between *hardware* and *software*. In the 1940s there was but hardware, which you had to tinker with yourself for the machine to do your bidding. With the advent of the stored program, the computing profession was divided into two: the hardware engineers who design the machine itself at the nuts-and-bolts level, and the computer programmers— the software guys—who appease the computer's insatiable appetite with a plethora of programs. Hardware is thus the machine itself, the conglomerate of plastic, metal, and silicon that you can touch and break. Software, by contrast, is somewhat of an elusive, ethereal material: It is the sequence of instructions that makes up your program. Software has in fact no material existence in and of itself: It only comes to be when placed within the cushy womb of a computer; its existence depends on the hardware substrate.

With general-purpose computers we did have to acquiesce, though, to one important drawback: Our computer is no longer the fastest gun in town. Most of the time, our primary goal as programmers is speed: We want the program to run as fast as possible since nobody likes the World Wide Wait. And the best way to do this is to construct a specialized machine—special hardware to do the task at hand (just like back in the 1940s). Then why has general-purpose computing supplanted custom-made hardware with such a vengeance? Our automatic cooking device can bake only apple pies, but it's one mean machine at that; however, it can do nothing else. The cook, on the other hand, is slower, perhaps less efficient, but can cook anything from lemon pies to seven-course meals, provided he is in possession of the right recipe. The situation is similar in the computing world: Building computer hardware whose purpose is to solve but one specific task is both a long and (very) expensive process; that is why it is rarely used, except for very specific niches. For example, an inexpensive telephone might have some special hardware for memorizing and retrieving ten phone numbers. The cost of building the specialized processor that handles this task may be quite high, but it will result in a very fast system; more important, the high design costs will be amortized over the millions of telephone units sold. By and large, though, computers today are general purpose: While you, the programmer, cannot change the hardware at all, you can, however, place within the computer's memory any imaginable program you can come up with, as long as it's written in a correct manner, using only operations that are understood by the processor. In short, we've opted for computers that are jacks-of-all-trades: Rather than being sensational at doing one task, they can do a good job on a wide range of tasks. This all boils down to that most fundamental force in the known universe: money. General-purpose computing is simply much cheaper.

During the past few years, a wave of nostalgia has been sweeping over parts of the computing industry. Computing professionals reminisce about the old days when you could change your computer's hardware, adapting the machine's entrails to suit your purpose, ultimately creating a system that puts all its competitors to shame. They created a new technology known as *configurable computing*, which is based on so-called configurable processors. Nostalgia, though, is not the driving force behind this novelty; we do not want to become once more switchboard operators and car mechanics. But—and this is the crux of the matter—we do want to enjoy the best of all possible worlds: the programming ease afforded by general-purpose computing along with the speed and efficiency benefits entailed by specialized hardware.

What is a configurable processor? Remember that the general-purpose computer has two main parts: processor and memory. The processor's hardware was designed by engineers, built in the factory, and from then on it can no longer be modified. All you can do is feed it with software—programs that are placed in memory. Wouldn't it be wonderful if you—the programmer—could modify the processor itself so that it would be much better adapted to the program it's supposed to run? Well, with configurable processors you can.

A configurable processor is actually quite similar to the general-purpose computer we've been talking about: You have your processor and memory—just as before. However, as opposed to the classical processor, the configurable one is *malleable*. The difference is in what goes into the memory: Whereas previously you placed a program in there, now you place a *hardware configuration*. As we mentioned earlier, the program placed in a general-purpose computer is a sequence of instructions that are chosen from a small set of allowable ones, recognized by the processor at hand. When we place the program in memory, the processor itself does not change by even the slightest amount. It's still exactly the same cook, but simply with a different recipe. What we place within the memory of a configurable processor, though, would best be called a hardware configuration rather than a program (perhaps we should also speak here of a *configurer* rather than of a programmer). This hardware configuration changes the processor itself. It contains a description of how the processor should rewire itself so as to execute the desired task. It's as if we had our automatic cooking machine rewire itself (at the hardware level) so as to make chocolate chip cookies rather than apple pies.

The whole thing sounds a bit suspicious, though: After all, once you've bought this chip—and it's sitting tranquilly on your desk—how can you dive in and change it? The answer is you don't. All you have to do is find the proper hardware configuration. Your job as a configurer is quite similar, in fact, to that of a programmer, albeit at a different level: Rather than find the sequence of instructions (the program), you need to find out how to wire the processor (the configuration).

What does such a configurable processor, or chip, look like in reality? Picture a large chessboard (with many more squares than the standard 8 × 8 one) on which each square represents a small unit of computation. In fact, each square is a tiny processor, albeit one that can do very little in and of itself. Forget adding and multiplying numbers; this little square can hardly accomplish such complex feats. So what *can* it do? First off, each square can "talk" with its immediate neighbors, the eight squares surrounding it. This means it can exchange bits and pieces of information with its neighbors—

indeed, a small, merry community. Well, not exactly. Actually each square can decide with whom it wishes to talk and whom it wishes to ignore. Assuming a somewhat less anthropomorphic stance, the *connections* between squares, that is, who communicates with whom, is one of the liberties you the configurer are given. It's up to you to design the so-called wiring scheme (for example, you might decide that one square will exchange information only with the neighbor to the north, while another will communicate with its three neighbors to the north, south, and east).

The fact that each square of our chessboard can communicate with its neighbors is definitely a good start; if they don't talk to one another, then nothing much will go on. What else can a square do? It can do very simple calculations known as *logic operations.* As opposed to a real chessboard, the squares are not permanently painted black or white: While at a given moment they assume either color, this may change at the next moment because of those logic operations. For example, a square may politely inquire of its good neighbor to the north as to its color. Armed with this knowledge the tiny processing square then inspects its own color and follows a simple *rule* such as "If both its color *and* that of its north neighbor are black, the square becomes black, otherwise it becomes white." In computing parlance this is known as the *and* logic operation and only a few additional such operations are needed to perform any complex computation. Many such simple rules are available. You can probably dream up a few with no problem. That's all there is to those mysterious logic operations. You, the configurer, must decide upon the rule that each square will follow.

How can these tiny processors—with such simple color-changing rules—lead to anything as complex as we see on our desktop computer? In fact, not only configurable processors work this way, but any processor on the market today is based—at the most fundamental level—on such simple operations. The point is that there are *lots* of squares (numbering in the millions). Force is in numbers.

Your job as configurer thus adds up to deciding how to *connect* the squares among themselves and what kind of simple (logic) *rule* each square will follow. This is ultimately expressed in the form of a hardware configuration, which you then place in the memory of the configurable processor. Now it's your turn to sit back and relax, and it's the processor's turn to rewire itself according to your wishes and start running that all-important job of yours.

Configurable computing thus succeeds in softening hardware, combining the advantages of both software and hardware. You can program (or configure) them just as you do with software, that is, they are general purpose, and this programming is done at the hardware level, so that you end up with a specialized machine that will swiftly carry out its mission. We have coupled the malleability of software with the speed and efficiency of hardware.

To date, classical processors are still not at risk of being supplanted; after all, they have had a head start of about 40 years. The latest ones to come out of chip companies such as Intel and Motorola are so fast because of advances in silicon-chip technology, so that even without their hardware being modified, they accomplish astonishing feats. Configurable processors, though, have begun to gain the computing industry's attention, proving themselves worthwhile in a number of important domains. For example—coming full circle back to military applications—suppose you want to build a computer to recognize flying targets such as aircraft and missiles.

Obviously, recognition must be achieved as fast as possible. A configurable processor could search the skies, make a preliminary, rough classification of the flying object, and then quickly rewire itself to become more adept at identifying what's coming. (If the object seems more like a plane, then the system would quickly rewire itself to act as a fast, efficient plane-recognition system.)

My colleagues Jacques-Olivier Haenni and Eduardo Sanchez sought to bring the power of configurable processors to the Internet, by constructing a *configurable network computer*. Over the past few years the Java™ programming language has gained ubiquity in the World Wide Web. Java underlies the recent proliferation of Web sites that are highly interactive. This computer language enables you to reserve your next vacation online, by viewing animated scenes (often in real time), clicking and choosing various options (single room? ocean view?), and finally typing in your credit-card information. Before the appearance of Java, Web pages contained only static information: text and images. When you accessed such a page on the Internet, your browser would copy its contents to your computer and display them. Nowadays, your browser copies computer programs written in Java and then runs them on your computer. Completely unbeknown to you, this is how you can reserve that long-awaited vacation: by accessing the Java-enhanced Web site of your travel agency. Haenni and Sanchez took this idea one step further: If you can copy (Java) computer programs and have them run on your computer, then why not transfer hardware configurations? You could then have a bare-bones (configurable) computer connected to the Internet, which would configure itself on the fly in accordance with your wishes. Haenni and Sanchez built a prototype of such a configurable network computer, an idea that has many interesting prospects: Imagine buying a very cheap computer (bare bones), plugging it into the Internet, and then having the computer modify its hardware to fit your needs.

The malleability of configurable processors means that they can *adapt* at the hardware level. Does the term *adaptation* ring a bell? Over the past few years, researchers working on bio-inspired systems have begun using configurable processors; after all, they reasoned, what could be better suited to implement adaptive systems than soft hardware? And what do you get when you marry configurable processors with evolutionary computation? Evolvable hardware.

In 1995 Eduardo Sanchez and Marco Tomassini from the Swiss Federal Institute of Technology organized in Lausanne, Switzerland what was to become the official inauguration of the field of evolvable hardware. In the preface to the conference proceedings, entitled "Towards Evolvable Hardware," they wrote: "We have crossed a technological barrier, beyond which we no longer need content ourselves with traditional approaches to engineering design; rather, we can now evolve machines to attain the desired behavior." The ability to cross this "technological barrier" is a result of advances in both evolutionary computation and in configurable computing. Sanchez and Tomassini went on to state boldly that "we are witnessing the nascence of a new era, in which the terms 'adaptation' and 'design' will no longer represent opposing concepts." Evolvable hardware aims to design machines that are adaptive, autonomous, and fault-tolerant.

Suppose you wish to build that plane-recognition system using evolvable hardware. First, you buy an off-the-shelf configurable device, most probably a so-called *field-programmable gate array* or *FPGA*. The "gate array" part means that it's a chessboard of configurable squares, like the ones we talked about earlier. "Field-programmable" means that the device can be configured (and reconfigured) in the "field," that is, on your desktop, as opposed to being configured once and for all in the factory.

To configure the FPGA to recognize planes, you need to provide it with an appropriate hardware configuration, one that will transform this blank-slate device into a plane recognizer. Finding the hardware configuration is usually quite hard, akin to the difficulty of programming standard computers. So let's do for hardware configurations what we did for computer programs: evolve them. That's the essence of this marriage: With evolvable hardware you apply evolutionary computation to evolve hardware devices (for example, evolving a hardware configuration for a plane recognizer).

How does this work in practice? There are two basic approaches in the field of evolvable hardware, known as *offline* and *online*. The distinction between the two is based on how much of the evolutionary process takes place in the configurable processor itself and how much of the process runs on a standard desktop computer connected to the processor via a cable.

The basic ingredients of simulated evolution are a population of genomes (candidate solutions to the problem at hand), fitness evaluation, selection of the fittest, and the genetic operators of crossover and mutation. With offline evolvable hardware, the configurable processor is merely a "slave" used for measuring fitness, while the evolutionary algorithm runs on a "master" computer. In a typical offline application within the field of adaptive robotics, Adrian Thompson from the University of Sussex in Brighton evolved a configurable processor that controlled a robot (a so-called controller); the goal was to move the robot within a rectangular arena of 2.9 × 4.2 meters without bumping into walls. The setup consisted of only one robot, which was used to test a given controller specification, that is, a given individual genome. (Thompson's robot contained a configurable processor, thus differing from Mondada and Floreano's Khepera from Chapter 4, which used a classical microprocessor.) If there's but a single robot, where's the population that we need for our evolutionary scenario? The population of genomes (candidate robot controllers) was kept offline on a standard personal computer connected to the robot. Each fitness evaluation involved a three-phase exchange between the computer and the robot:

1. The computer places the genome to be evaluated (a hardware configuration for a robot controller) within the (single) robot, thus configuring it to behave in the manner specified by the genome.
2. The robot is allowed to run about the arena, while the computer collects statistics on its performance (how fast it is moving, how many times it bumps into walls).
3. Based on these statistics, the computer computes the robot's fitness score.

With offline evolvable hardware, the population is virtual. Thompson had thirty virtual robots, which were all stored in the computer. This can be done since a robot is specified entirely by the hardware configuration, which is nothing more than bits and bytes, a long string of 0s and 1s that the computer can store within its memory. Except for fitness evaluation, which is done using the real robot, the entire evolutionary process is simulated within the computer. The population is virtual and so too are selection, crossover, and mutation. Once a generation of (thirty) robots has been evaluated one by one on that single robot we have, the computer has a fitness score for every genome. It can now select the fittest, cross them over, and sprinkle mutations here and there to create the next generation of robot hardware controllers.

Whereas offline evolution requires an external, supervisory computer, with online evolvable hardware the entire evolutionary process is embedded within the configurable processor. Thus, maintaining the population, individual fitness evaluations, selection, and the genetic operators of crossover and mutation are all carried out online, that is, onboard the processor. To my knowledge, the first online evolvable device was the *Firefly machine*, constructed in 1996 by my colleagues and me at the Swiss Federal Institute of Technology in Lausanne.

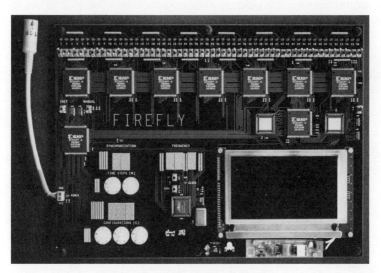

Figure 6-1   The Firefly machine
COURTESY OF ANDRÉ ("CHICO") BADERTSCHER

Fireflies in nature are fascinating creatures: Thousands of them will swarm together at night, their corporal lights flickering in a chaotic manner. Then, a few minutes later, they all flash on and off in unison like a well-trained troop of soldiers. Unlike soldiers, though, there is no general involved. Somehow, the multitudinous interactions taking place on a local level, with each firefly observing its nearby neighbors, end up in global synchronization, the entire group flickering in unison.

The Firefly machine contains a small troop of digital fireflies. At the top you can see a row of lamps (fifty-six in all), each of which represents one digital firefly. These lamps can be either on or off. As with their natural kin, the digital fireflies start out flickering chaotically and *evolve* to flicker in unison via an online evolutionary process. Online means that every aspect of the evolutionary process—population, fitness evaluation, selection, crossover, and mutation—is on the board itself. Evolution takes place within the FPGA chips you see in Figure 6-1: the row of seven black squares below the lamps. Because it is online, there is no additional computer; what you see in the picture is all there is. (The screen is used to display fitness scores and evolved genomes, and the knobs to the left are used to set some parameters of the evolutionary process.)

The Firefly machine is neatly tucked away in a gray office cabinet. From time to time, especially when a visitor says "please," we take it out, switch it on, and stand back: After all, it does evolve all by itself, online. Blinking lights are nice, you may think, but does Firefly serve any useful purpose? The answer is, No; it was not intended to. When we built it a few years ago, at the dawn of the evolvable-hardware age, we had a humble goal in mind: to build a proof-of-concept of the idea of online evolvable hardware.

More recently, Tetsuya Higuchi and his colleagues at the Evolvable Systems Laboratory of the Electrotechnical Laboratory in Tsukuba, Japan have been designing and fabricating several online evolvable chips for commercial applications. One of these is a myoelectric prosthetic-hand controller. The myoelectric hand is operated by the signals generated with muscular movement (electromyography, or EMG). A disabled person usually requires over a month of training to be able to manipulate with ease a prosthetic hand. Rather than have the person adapt to the hand, Higuchi and his colleagues reversed this scenario by having the artificial hand adapt itself to the disabled person. The online controller evolves to map EMG signals to desired hand actions (there are three pairs of actions: open-grasp, supination-pronation, and flexion-extension). Because EMG signals vary greatly between individuals, you cannot have the controller come preprogrammed from the factory. With the evolvable-hardware controller, the hand usually requires less than 10 minutes to adapt to its owner through online training, in which the person repeats various hand movements. This is a notable improvement over the 1 month required when the designer is the one doing the adapting.

The online approach allows the owner to embed the evolving device in the target environment. The hand-controller chip has to be placed in a lightweight prosthetic hand. In an online robotic application (as opposed to Thompson's offline experiment) we wish to rid ourselves of the cumbersome desktop computer so that the robot may roam about with an internal evolvable chip. Some form of online adaptation is, in fact, the *only* possibility where dynamic environments are concerned: A robot sent to explore the surface of Jupiter cannot be preprogrammed entirely in advance and will therefore have to adapt online.

Configurable computing will probably continue to advance in leaps and bounds over the next few years. One possible development will be its merging with the classical-processor industry, seeing the construction of processors that are partly hardwired, partly configurable. Ultimately, such soft hardware might end up being quite ubiquitous, with your desktop computer able to adapt to the whims and fancies of the world at large.

Sometimes, being a softy isn't so bad after all.

CHAPTER 7

A Stitch in Time Saves Nine

One of the most ancient folk wisdoms is that of time as a healer of wounds. Isn't that what your mother used to say to you—"in time it'll heal"—when you'd come home full of bruises after that rough-and-tumble game with the kids from the next block? The ability to heal is one of the prime characteristics of living beings, whereas machines exhibit a rather regrettable tendency simply to break down. (This is seen in the archetypal "damn-this-car" scene, where you kick the carcass of your dilapidated automobile—which has just broken down on your way to that important job interview—followed by the inescapable ride to the garage with an annoyingly cheerful tow-truck driver.)

You've probably guessed what comes next: machines that heal, or—to be somewhat less anthropomorphic—self-repair. In his book *The Logic of Life* the Nobel laureate, François Jacob (based upon an argument given by the philosopher Immanuel Kant), affirms that: "A watch can neither produce the parts that are removed from it, nor correct their defects by the intervention of other parts, nor rectify itself when it loses the correct time." While this may have been true when Jacob's manuscript was published in 1970, this statement is no longer true. There is at least one watch that can self-repair: the BioWatch™.

The BioWatch was designed by a group of researchers from Switzerland, headed by Daniel Mange from the Swiss Federal Institute of Technology in Lausanne and by Pierre Marchal from the Centre Suisse d'Electronique et de Microtechnique in Neuchâtel, as part of an ongoing project known as "Embryonic Electronics," whose ultimate goal is the construction of integrated circuits (computer chips) that are able to self-repair. We already saw how evolution and learning serve as inspiration for evolutionary computation and artificial neural networks. The mechanisms underlying the workings of the BioWatch are inspired by yet another biological process known as *ontogeny*.

A human being consists of approximately sixty trillion (sixty followed by twelve zeros) cells. At every moment, and in each of these cells, the three-billion-nucleotide genome is decoded to produce the proteins needed for the survival of the organism. This genome contains the individual's genetic inheritance, and, at the same time, the instructions for both the construction and the operation of the organism. The parallel execution of sixty trillion genomes in as many cells occurs ceaselessly throughout the individual's lifetime. Faults are rare and, in the majority of cases, successfully detected and repaired.

The process by which such complex organisms come to be is that of ontogeny: the development of a multicellular being, possibly containing trillions of cells, from one single mother cell—the zygote. Ontogeny involves two basic sub-processes: cellular division and cellular differentiation. The fertilized mother cell (the zygote) divides successively, the one-cell organism becoming a two-cell one, and then a four-cell one, and so on. A crucial aspect of this division process is that each newly formed cell possesses a copy of the original genome, the recipe used to build *the entire organism.* Cellular division is accompanied by a process of differentiation: As the daughter cells (the new copies) are formed, they *specialize* in accordance with their surroundings—their position within the growing ensemble. The ontogeny of animals, and especially of humans, is usually referred to as embryogenesis, involving the formation and development of an embryo.

What does all this have to do with computer chips? Enter Embryonic Electronics. Imagine a chessboard, albeit one with many more squares than the familiar 8×8 one, where each square is a small, simple computer; much simpler, in fact, than the one humming on your desk. Does this sound familiar? Indeed, the actual hardware used to implement the "organisms" described herein is the configurable chip of the previous chapter. Biologically minded as we are, we refer to each such square as a *cell.* Our goal is to create multicellular artificial organisms whose purpose is to exhibit a specific behavior defined by us. For example, one such organism, composed of several squares—or cells (hence *multi*cellular)— might be designed to exhibit the time of day, thus functioning as a watch.

Let's consider this simple watch (BioWatch) example (see Figure 7-1). It is an organism composed of four adjacent cells aligned in the same row; the two left ones count hours (from 00 to 23—since there are two digits, we need two cells) and the two right ones count minutes (from 00 to 59—again two cells are needed). (Note that the BioWatch uses but one dimen-

Figure 7-1   The BioWatch
COURTESY OF ALAIN HERZOG

sion—one row of the two-dimensional chessboard; more complex organisms make use of both dimensions.) This watch organism is defined by a recipe, or *genome*, that contains a sequence of instructions (or *genes*) for building the watch. Now comes a crucial point: Each cell of this four-cell organism contains the *entire* genome, and not just the specific genes that are actually used by the cell to carry out its intended function. This is analogous to Nature: For example, in the human body each cell contains the entire genetic makeup (all three billion nucleotides), though it uses only specific genes, depending on the cell's functionality—whether it is a liver cell, a kidney cell, and so on. How does each cell "know" what it's supposed to be? Well, let's not get ahead of ourselves.

How does the watch organism come to be? First off, it must be understood that we assume the existence of a large chessboard of blank-slate cells. To wit, as opposed to Nature, no material is actually created; it's given to us a priori, but in a *pristine* state. Our job is to turn this inert chunk of material into a functioning device—or organism. One possibility would be simply to place the organism's genome (designed by us, the human engineers) in each of the four cells. This solution would work nicely for such a small organism, but it would get more and more cumbersome as the number of cells grew; when considering billion-cell organisms or larger, this simple "hand placement" strategy simply won't do.

Nature, though, has found another solution: cellular division. She sets out with just one cell, containing one genome (this is known as the zygote). This cell then divides into two, resulting in an embryo that is made up of two cells, each with a complete copy of the genome. One more step, where each of the two new cells further divides into two, brings about a four-cell embryo. This process is extremely rapid: After just twenty divisions the embryo already contains one million cells.

Coming back to our artificial organism, we now use this cellular-division "trick." All we have to do is place a "zygote"—one copy of our designed watch genome—at the bottom-leftmost cell of the chessboard. (Since the watch is one-dimensional, we simply refer to this as the leftmost cell.) Then we stand back and let this genome duplicate itself into the other three cells to the right. With two-dimensional organisms, the effect is even more spectacular. If we place the original zygote in, say, the middle cell, then—like ripples in a pond after a stone has been thrown in—it will divide to fill the chessboard region prescribed by the genome.

With cellular division over, each of the organism's four cells is in possession of the genome. But they all possess the *same* genome: How, then, does each one zero in on its functionality? After all, though their genetic material is identical, their respective tasks are definitely not (for example, the left hours cell counts up to 2, whereas the left minutes cell counts up to 5). This is quite similar to nature, where liver cells and kidney cells have the same genomes but do entirely different things. It's time to differentiate.

Cellular differentiation in nature is that process wherein each cell assumes its intended role by extracting the relevant instructions from the genome. In *The Triumph of the Embryo*, the embryology expert Lewis Wolpert wrote on differentiation: "There are, then, two quite difficult things that cells have to do. The first is to know where they are, that is, acquire positional information, and the second is to use this information appropriately." While nature may work hard to implement this scheme, our artificial organism presents a much easier case. (Incidentally, one of the major success stories of biology over the past two decades has to do with discoveries concerning embryonic development in Nature, and in particular the use of chemical gradients and homeotic genes to acquire positional information and to employ it appropriately. This recently spurred Matt Ridley to write: "Indeed, so simple is embryonic development in principle— though not in detail—that it is tempting to wonder if human engineers should not try to copy it, and invent self-assembling machines." An interesting piece of insight, indeed, if unknowingly hindsight.)

Remember that our organism is implemented using very simple computers—squares of a configurable chip—arranged in a chessboard pattern Each such square can potentially come to be "inhabited" by a genome during cellular division, in which case it becomes part of the multicellular organism. Each cell can therefore "acquire positional information" in quite a straightforward manner, simply by finding out its precise location within the board—its row and column numbers. There is a small catch, though: This positional information is not given at the outset; it is not "engraved" a priori into each of the blank squares. Why is that? After all, what could be simpler than etching onto each square of the board its absolute "address"? The reason for not doing so is that each cell must know its *relative* position *within the organism* and not its absolute coordinates on the board. The watch organism we are discussing here is but a simple example (not, in fact, even two-dimensional, using only one row of the board). However, the ultimate goal is to use huge chessboards—with billions of cells—which will thus contain *several* organisms. Thus, we must ensure that each cell knows its relative position within the organism to which it belongs.

During cellular division of the watch organism, we first placed the original genome (the zygote) in the leftmost cell of the chessboard, after which this genome duplicated itself into the other three cells. Let's call this leftmost square "cell 1." When the genome duplicates itself into the cell to the right, this new cell then takes a peek at its neighbor's coordinate, increments it by one, and thus gleefully establishes itself as cell 2. Continuing in this manner, the process is thus quite simple. When genome duplication (division) occurs, the receiver cell increments by one the position of its left neighbor. What happens when we're dealing with a two-dimensional organism? In this case we need to manage *two* positional coordinates instead of one: row and column. Placing the original genome in the bottom-left cell, it would then duplicate itself both rightward and upward. When moving upward, we increment the row counter by one, and, when moving rightward, we increment the column counter by one. For example, when the genome arrives at the cell that is one square up and two squares to the right, then the acquired coordinates will be: row 2 (from the bottom), column 3 (from the left).

Our multicellular watch organism is now almost ready for action: Each of its four cells possesses the genome, and each knows its position within the organism. To complete the cellular differentiation phase, each cell *extracts* from the genome the gene (or genes) that defines its own functioning—in accordance with its position. Since each cell has a different coordinate within the organism, it will extract a different set of genes, a distinct set of instructions to guide its behavior. Thus, the leftmost cell will come to count tens of hours, the cell to its right will count hours, the third cell will count tens of minutes, and the rightmost cell will count minutes.

Cellular division and cellular differentiation have done their job. The watch organism is now primed for action and will carry out its function in a precise, Swiss fashion, showing the time of day.

Why construct such a BioWatch? While this kind of embryonic process is admittedly fascinating, with all these intricate developmental phases, there are much simpler ways to build watches. This brings us to the crux of the matter: Stitching time; our BioWatch can heal upon suffering injury. There it is—our four-cell BioWatch—cheerfully carrying out its (timely) mission in life when suddenly disaster befalls: One of the cells breaks down. A cell is physically a silicon device, and these do exhibit an unfortunate disposition toward malfunctioning from time to time. In order to repair the damage, the organism must be capable of doing two things. First off, it should be able to *detect* the fact that something bad has just happened, after which it can try to *repair* the damage. We won't talk about the first part—fault detection—which is something that electrical engineers know how to do (in their jargon the name of the beast is BIST: Built-In Self Test). Let's see how the organism can make repairs once it *knows* that something's amiss.

Our four-cell BioWatch inhabits but *part* of the chessboard universe. In other parts of the universe there may be other organisms, carrying out different tasks, and—luckily for the BioWatch—there are also some "spare cells," blank squares that have not been used up until now. For simplicity, let's suppose that there's one such spare cell just to the right of the BioWatch's rightmost cell; let's assume, furthermore, that it is cell 2 (second from the left) that has broken down. The BioWatch has detected that cell 2 is no longer functional, and declares it "dead." The genome is then copied from cell 4 into the new spare cell, cell 5. The situation is now as follows: The four-cell organism is spread over cells 1, 3, 4, and 5, with cell 2 considered as dead, unused tissue. But how does the watch continue to function correctly? After all, cell 3 must now function as the *second* cell; that is, it no longer counts tens of minutes, but rather must start to function as the hours counter.

The secret is in the positional information—the heart of cellular differentiation—which is relative and not absolute. After the organism works its way around the dead cell, the cellular coordinates are recalculated: Cells 3, 4, and 5 acquire new coordinates—2, 3, and 4, respectively. Do these look familiar? Lo and behold, they are the exact cellular coordinates of the original, "healthy" organism. The end result of the repair process is that the four-cell organism occupies four cells whose coordinates are 1, 2, 3, and 4, with a nonfunctional cell—dead "scar" tissue—between cells 1 and 2.

The organism's composition as a multicellular structure, with each cell possessing the entire genome (and not just its own genes), allows the cellular processes described above to take place: cellular division, cellular differentiation, and cellular repair. The repair process of the BioWatch not only overcomes the damage, but does so without losing the correct time! Thus, it not only heals in time, but heals time itself. (As far is the actual implementation is concerned, the ability to retain the correct time throughout the repair process is achieved in a BioWatch version that is slightly more complex than the one described herein; the basic ideas are nonetheless the same.)

Notice that when a cell is found to be defective, it is immediately declared as "dead," to be regarded henceforth as unused—and unusable—scar tissue. If several faults occur in succession, then the chessboard may end up looking like a scorched battlefield, laden with the carcasses of dead cells. Can this be avoided? The Swiss BioWatch-makers responded in the affirmative by noting that cells in Nature are capable of quite astonishing feats. Embryologist Lewis Wolpert wrote on this issue: "Cells are more complicated than embryos. Nothing illustrates this better than the remarkable capacity of single-celled organisms to regulate the pattern on their surface in a manner very similar to that shown by regenerating multicellular animals. Understanding this process could reveal a fundamental developmental principle." The Swiss team thus decided to imbue the cell with *internal* repair mechanisms that would avoid the need to discard it as soon as it broke down. After all, maybe it's just a tiny fault. Why throw away an (almost) perfectly good cell?

How can you fix the cell on the inside? The basic idea is to divide each cell into yet finer elements, dubbed *molecules*. Whereas before we had only spare parts at the cell level— spare cells on the chessboard—we now maintain spare molecules within each cell. When a cell breaks down, the organism no longer calls immediately for a hearse but rather for an ambulance. The damaged cell's internal repair mechanisms kick into action, and, provided there are a sufficient number of reserve molecules, a healthy cell, fit as a fiddle, will soon emerge to relentlessly count those precious minutes. Only when the supply of spare molecules has been exhausted will the cell receive its last honors, and the repair process will take place at the cell level, by requisitioning an unused, spare cell on the chessboard.

Our system has thus become *hierarchical*, functioning on two distinct yet interacting levels: the chessboard of squares (cells), each of which is itself a chessboard of molecules. This renders the repair process much more efficient since one can first attempt internal cellular repair at the molecule level before deciding that the damaged cell is gone for good, whereupon it's time to delve into the reservoir of unused cells. Increased healing capabilities are not the only advantage, though, of this molecular-cellular system. It is also much more flexible: We can "custom tailor" not only the organism's behavior but also its structure.

In the single-level system, the genome contains genes (instructions) for the behavior of each cell; in the example of the BioWatch, the genes specify the manner in which the cell should count minutes or hours. In the two-level, molecular-cellular system, there is an additional part in the genome that specifies the dimensions of the cells, that is, how many molecules make up a cell. This means that different organisms not only can have different intended functions, but indeed their cellular structure may be altogether different. This allows for the design of organisms that make much better use of the resource at hand—the electronic circuit.

Obtaining self-repairing machines that make economical use of resources is perhaps the major motivation for constructing such "embryonic" organisms. A classical repair method in engineering—used in mission-critical systems such as spacecraft—is to simply multiply each subsystem by three, and then carry out a "majority vote." Before executing an important action, each of the three clones "voices" its opinion;

obviously, if all three are in order then they will advocate the exact same action. However, if one has failed, then the other two will catch on to this, since they'll vote for the same (correct) action, thus constituting a winning majority (ensuring the functioning of the system as a whole, for example, the continued voyage of the spacecraft toward Jupiter). Such a scheme can only tolerate a malfunction in one of the three cloned subsystems. Two failed subsystems mean that there is no majority (since each clone votes differently), or, worse yet, that the wrong majority decision is made (if both failed subsystems exhibit the same malfunction). In order to tolerate two malfunctions, we'd need to add yet more cloned subsystems.

Embryonic-style electronics, based upon the molecular-cellular structure described herein, results in much less costly machines, by eschewing this multiplicity-of-subsystems problem. Essentially, what we're doing here is carrying out repair at a much finer level: Instead of chucking out an entire component, we can hone in on the very small malfunctioning molecule or cell and fix it. While spare parts are still needed (both molecules in each cell, as well as cells on the chessboard), the total cost is drastically reduced as compared with multiplying the whole shebang by three.

While the Swiss BioWatch designers have their feet planted firmly on the ground, their heads have lately found their way to the clouds. They have designed the "giant BioWatch," a room-sized BioWatch that they plan to display at a major exhibition in Switzerland. The giant BioWatch contains thousands of molecules and will allow visitors to "kill" components interactively (molecules and cells). They can then stand back and observe the flickering lights as the watch heals itself before their eyes. This watch is shown in Figure 7-2.

Figure 7-2   The Giant BioWatch
BY ANNE RENAUD AND REPRODUCED BY HER KIND PERMISSION

Silicon circuits can behave like carbon beings: Plant a seed—the embryo—and watch it develop and grow into a full-fledged organism. Being multicellular, and with each cell in possession of the entire genome, you can then sit back and relax: Harm will come—and go.

"If you can look into the seeds of time, And say which grain will grow and which will not," wrote Shakespeare in *Macbeth*. Where the BioWatch is concerned, the seeds of time definitely grow.

CHAPTER 8

Not Only Diplomats Have Immunity

Slowly the invasion begins. Numbering in the millions, the invaders are determined to destroy their enemy. The enemy, however, is not quite ready to throw in the towel and has agents spread throughout its territory, which seek continually to detect the offensive foreigners. These agents cling to their targets, pinning them down until the arrival of the death squads, which annihilate the invaders.

Storm troopers attacking a rebel planet? Klingons invading Federation space? Not exactly. The war in question is the one that takes place in your body when you've come down with the flu. Our bodies are equipped with a wonderful defense mechanism, evolved over the eons, known as the *immune system.* Its job is to keep us up and running about in shipshape condition, warding off foreign, harmful elements that enter our bodies. These elements go by the name of *antigens* and include all kinds of nasty little beasts, such as viruses, bacteria, and parasites. When these meanies enter our body, the immune system springs into action. Detector agents—T cells, B cells, and antibodies—hunt down the foreigners and physically bind to them. The death squads—macrophage cells—then come along and finish the job, eliminating the malevolent antigens.

The immune system of living beings exhibits a number of striking characteristics. First off, though conducting a war, it is not organized like an army. There is no chain of command, ending with the top brass; rather, the system is entirely distributed, with no centralized control that initiates or manages an immune response. The immune system's success arises from numerous localized interactions between its agents and the foreign antigens. Having no central headquarters has the advantage that the system can tolerate many kinds of failures.

Another important property of the immune system is its ability to detect previously unseen foreign material. After all, this system had evolved long before the arrival of modern medicine and could not rely on our being vaccinated against diseases; the system must be able to cope with new, unknown enemies. When a novel infection sets in, the immune system initiates what immunologists call a primary response, creating new detectors specialized for this hitherto unseen disease. The system thus *adapts* to new situations. Moreover, it also *remembers* previous infections. When you come down with the same disease (or some similar variant of it) for the second time, the immune system mounts an aggressive attack known as a secondary response, usually resulting in a more speedy recovery. In fact, sometimes, you don't even know that a battle has just been fought—and won!

During your lifetime, your immune system learns by implementing Friedrich Nietzsche's maxim: "That which does not kill us makes us stronger." Born with the basic options in place, your encounters with the enemy then serve to add helpful accessories (notice how a child is easy prey for disease, her immune system still being in its infancy). These encounters may be haphazard or, as with common vaccinations such as that against smallpox, intentional. The latter is based on the idea of pitting your body against the enemy, albeit a weaker one at that. Thus, the immune system engages in limited warfare (the primary response), during which it learns much about the enemy. If ever the enemy dares attack a second time (even if it does so in great force), the immune system will launch a formidable secondary response and have you back on your feet in no time.

Since we have been exposed to distinct foreign elements during our lifetimes, your immune system and mine are different; our two systems have gone to different schools. The immune system's ability to learn is partly why some people seem more resilient in the face of maladies than others. (There are also genetic causes; for example, there are rare cases of individuals who are born with no immune defenses at all, to whom even a common cold may be fatal.)

As we saw in the previous chapter, mechanisms that underlie the basic structure of multicellular organisms can be instrumental in bringing about self-repair. Once organisms grew more complex, though, these mechanisms turned out to be insufficient, and Nature invented an entirely new repair mechanism, the immune system. Yet another of her tricks for keeping complex machines (such as mammals) up and about without their falling prey to every marauding invader.

Just as animals need protection from foreign evildoers, so do modern-day computers. In the early days of computing, there were few computers, and they were treated like an endangered species, with access to them being strictly controlled. Evil-minded actions were thus virtually unheard of. This has changed drastically over the past decade with the enormous increase in the number of computers worldwide, coupled with the remarkable rise of the Internet. For better or for worse, we're all connected nowadays. A diskette, a CD, or an email can be harbingers of good, bringing along the latest version of your favorite game or greetings from a long-lost friend, but they can also be messengers of evil and hide some new strain of computer virus. Such viruses can be anything from completely innocuous to downright deadly, the latter possibly wiping out every single iota of information stored in your computer.

Over the past few years, computer viruses have been highly successful in pervading our computers. What exactly are they? At heart, a computer virus is a *program*, a small creature that inhabits a world separate from ours, namely, the insides of a computer. As with their biological counterparts, computer viruses have one primary objective: to replicate. In most cases they also have a secondary goal—that of causing some observable effect, ranging from displaying a harmless message on the screen to completely erasing your disk.

A computer virus is born in the mind of a malevolent or ludic programmer who writes the first specimen (in the same way that a new virus is created in a biological-warfare laboratory). The virus replicates by adhering to a host and coopting the host's resources to make copies of itself. Whereas the natural virus's host is a biological cell, the computer virus's host is a computer or another program.

And so, having written the virus program, its creator now infects another program with it, preferably a popular one such as a commonly used word processor. No test tubes and messy procedures are involved here; all one need do is glue the virus program onto the beginning of the word processing program, thus creating an infected version of the latter. Now the creator places the infected word processor (or program) in an accessible place, for example, on diskettes that he exchanges with friends or foes, or on the Internet.

Then you come along, experiencing one of those days when everything seems to go wrong, when the entire world appears to be in cahoots against you. "At least I've finally got the latest version of my favorite word processor," you think to yourself. Well, it definitely *is* one of those days: The diskette contains an infected version of the program.

What happens when you run the infected word processing program? The computer sets about running the small viral program, which is attached to the beginning of its infected host program. The virus immediately entrenches itself within the computer's memory, creating a copy that is separate from the host program; thus, it can continue to do its creator's bidding even when the host is long gone (that is, when you have quit the word processing program). Next, the virus passes control back to the infected program, which proceeds to work normally. The delay caused by the copying of the virus is so small that at this stage it is virtually impossible to detect that anything out of the ordinary has happened. Tired of writing that important document for tomorrow's meeting, you quit the word processor and run some game software. As the computer runs this game program, the dormant virus, lurking within the computer's memory, springs into action: It inserts a copy of itself into the previously uninfected program. The cycle of virulence can thus repeat itself: An infected program inoculates a hitherto "healthy" computer with the virus, which then proceeds to infect every program that is run on that machine. Since all this takes place unbeknownst to the user, she will probably become an inadvertent agent who helps spread the epidemic by passing along infected programs.

In addition to attending to its own replication, a computer virus usually performs some additional task, which may be harmful. Clearly, we need to immunize our computers. Enter antivirus software.

Antivirus programs have existed since shortly after the appearance of the first computer viruses in the mid-1980s. How does such a program detect viruses? One approach is to monitor the computer for viruslike behavior, looking for suspicious-looking programs or ones that seem to have been modified. This approach might even discover hitherto unknown viruses; however, it might also produce false alarms since some legitimate activities resemble the workings of viruses.

Another approach is to scan the entire computer system (using a special scan program) for specific patterns—small pieces of programs—that are indicative of known viruses. While such an antivirus program rarely raises false alarms, it must, however, be continuously updated as new viral strains come into existence. Antivirus programs do their best not only to detect viruses but also to disinfect the computer, repairing as much of the damage as possible.

A good antivirus program should be both specific—able to detect only viruses, without raising false alarms—and comprehensive—able to detect as many known (and perhaps unknown) viruses as possible. This has required extensive work by computing experts, somewhat similar to the work of virologists: analyzing new virus strains and coming up with ways to defeat them. Since dozens of new viruses are introduced each week (most of which, happily, do not proliferate), antivirus specialists have their hands full. Quite naturally, they seek to automate this process, with some of the recent approaches based on natural immune systems.

Jeffrey Kephart, Gregory Sorkin, David Chess, and Steve White from IBM's research center in upstate New York proposed what they called an immune system for cyberspace, which automatically finds prescriptions for detecting and removing new virus strains. Their test system consists of a small network of personal computers, each of which runs a special antiviral program. This latter continuously stalks the computer's innards in search of suspicious activity (such as unaccounted-for program modifications or patterns associated with known viruses). When a program thought to be infected is found, the antiviral monitor sends a copy of it over the network to a central virus-analysis machine.

This machine is a computer that acts as an automatic laboratory for analyzing possibly infected samples of computer programs. It does so by creating a digital "Petri dish" in which various programs are run, thus luring the putative virus into doing its noxious work. Any programs that are infected in this Petri dish serve to extract information that will be used to identify and remove the virus in question. As with a real biological laboratory that deals with dangerous specimens (the *Ebola* virus, for example), the digital lab is also tightly sealed off from the rest of the (network) world.

Having automatically run the experiments, the analysis machine sends a complete description of its findings back to the infected computer, which then uses the information to combat the malicious invader. Moreover, all the other computers connected through the network also receive a copy of this information so that they may win their own battles in turn. The whole operation—identifying possible evildoers, sending them to be analyzed, and using the returned analysis results—takes place with no human intervention—an autonomous immune system for cyberspace indeed.

Immunologists usually describe the chief role of the immune system as that of distinguishing "self" from dangerous "nonself," and eliminating the latter. Stephanie Forrest, Steven Hofmeyr, and Anil Somayaji from the University of New Mexico have been designing computer immune systems that more closely resemble natural ones, especially concentrating on the self/nonself issue, on how to differentiate between "good" and "bad" behavior. Not an easy problem to solve as even natural immune systems can err by confusing good self with bad nonself to give rise to so-called autoimmune diseases.

Forrest, Hofmeyr, and Somayaji first sought an appropriate way to define "self" in a computer. As we've seen, a program is a sequence of instructions whose execution accomplishes a desired task such as word processing. The idea is therefore to build up a profile of normal behavior for each program of interest, consisting of a description of legitimate patterns of instruction sequences; programs that behave incongruously with their profiles are then considered suspect. The profile is specific not only to a particular program, but also to a particular computer and to a particular user (for example, I might use options of the word processor that are different from the ones you use, thus resulting in different behavior profiles). This means that your computer and mine will come to have different immune systems, in a manner reminiscent of how things work in Nature.

Another mechanism for immunizing computers, proposed by Forrest and her colleagues, is based on the process of T-cell maturation. T cells are an important class of detector cells in the immune system. When T cells are first created in the body, their binding behavior is random, and thus they may bind not only to antigens but also to self proteins—provoking an autoimmune response. In order to solve this problem, the immune system sends the newly formed T cells to an organ called the thymus to mature.

T-cell maturation involves several stages, one of which is known as negative selection: Self proteins constantly circulate through the thymus. If a T cell binds to any such self element, then it is destroyed; otherwise, it is allowed to mature, leave the thymus, and become part of the active immune system. Thus, only T cells that fail to bind to self proteins emerge from this sieve. These are potential nonself detectors since the proteins to which they *do* bind are most likely nonself ones.

Building a computer immune system based on T-cell maturation works as follows: Generate detectors at random and have them negatively selected; that is, test each one to see if it matches self patterns. The detectors as well as the self elements are simply small pieces of programs. Any detector that fails to match self is considered a nonself detector, which will be used thereupon to monitor the computer's activity. When such a detector is activated, it means that some suspicious nonself element has been discovered, whereupon an immune response will be set off.

An interesting point about this system is its use of imperfect detection. The match between detector and detected does not have to be perfect, but may be partial. Such approximate binding is another trick borrowed from natural immune systems: The bond between a T cell and an antigen does not have to be perfect in order to trigger the immune response that culminates in the antigen's destruction. This is quite advantageous since creating perfect detectors is quite hard; you can think of this as the difference between identifying Rusty the dog, and just recognizing that the animal in question is a dog.

In the years to come we will probably see a continual arms race between computer viruses and antiviral programs. Though research in the latter area is still in its infancy, it is quite exciting to see how computers acquire immune systems that take after natural ones. Still, as things stand now, it seems that the best way to immunize your computer completely is to place it in a diplomatic pouch—and leave it there untouched.

CHAPTER 9

Computing in a Test Tube

Within the purview of scientific research, the field of molecular biology has taken unarguably some of the greatest strides over the past few decades. We started out the twentieth century with but a skimpy sketch of the internal workings of our bodies. We've entered the twenty-first armed with a draft of the entire three-billion-nucleotide human genome: our genetic inheritance.

Every living being on Earth caches within each of its cells the instructions for its development, the *genome*, itself composed of *genes*, the elemental unit of heredity. In some cases one gene gives rise to exactly one trait (for example, in fruit flies, a defect in a single gene results in the fly having a leg growing out from the head in place of the antenna). In most cases, however, a trait is defined by an intricate interplay among an ensemble of genes; this is the case with many human higher-level functions, such as intelligence, creativity, sociability, and so on. Such traits are very difficult to study at the genetic level since many genes have their "say" on the matter. To complicate things even further, human beings are subjected to an environment after their birth; our behavior develops as a result of contact with various sources, such as family, friends, television, books, and so on. (This is also true, though to a lesser degree, of other animals, especially mammals.) The interplay between genetics and the environment usually goes by the catchphrase "Nature versus nurture."

What have biologists taught us over the years? Every organism on Earth (whether a *Streptobacillus* bacterium, an African elephant, or a giant sequoia) is defined by a "recipe": a list of instructions, the genome, which—when correctly decoded and executed—gives rise to the organism in question. In case of multicellular organisms (such as the elephant and the sequoia) each and every cell contains a copy of the genome.

The genome is basically a long list of letters, taken from a four-letter alphabet: A, C, G, and T (standing for adenine, cytosine, guanine, and thymine); biologists refer to these letters as *nucleotides* (as mentioned above, there are about three billion letters, or nucleotides, in humans). The nucleotides link up to form chains of deoxyribonucleic acid, better known as DNA. (One of the triumphs of modern molecular biology was the discovery in the 1950s of DNA's double-helical structure, the all-familiar intertwined pair of slivers that have by now become one of the icons of our times.) DNA is thus the molecular basis of heredity. Before ending this short excursion into molecular biology, let us introduce just one last term, *oligonucleotide*, which is a short chain of usually up to twenty nucleotides.

Genomes are thus composed of genes, and these latter are made of DNA. These structures are at the basis of all life: Your genome, obtained by combining genetic material from both your father and your mother, is what gave rise to the human being who is now reading these lines.

Does this have *anything* to do with computing? Much of the ongoing work in biology (such as decoding the human genome) would be either extremely difficult or downright impossible without the aid of substantial computing power. High-performance computers are routinely used these days by biologists to study many of the basic questions regarding *Life* and have, over the past few years, become standard laboratory tools. Nowadays, it's rare to find a scientist who does not make use of computers on a regular basis.

But can biology be used in computing? We saw how Nature can serve as a source of *inspiration* for building better computing machines that adapt, evolve, learn, and heal. But biology can also serve the cause of computing more directly, by using DNA molecules as building blocks for a novel type of computer. This field, launched by Leonard Adleman from the University of Southern California in 1994, is known as *DNA computing*.

Let's put biology aside for a moment and talk about a serious conundrum in computing science: the traveling salesperson problem. It is an archetypal example of a problem that underlies numerous real-world applications, yet is extremely hard to solve. Suppose you're an itinerant salesperson planning a tour of five cities by car. Your objective is to visit every city exactly once and then come back to your point of origin. Obviously, you want to plan the shortest possible route, making sure that the total distance covered is minimal. After all, you're paying for the gas, and besides, time is money.

How would you go about planning your itinerary? One way, known in computing parlance as the "brute-force" method, is to examine all possible tours. An itinerary is ultimately a simple list of five cities written in the precise order in which you will visit them. (That is, you start with the first city on the list, then travel your way down the list, ultimately ending up in the fifth city; we need not list the sixth city, which is implicitly assumed to be your point of departure.) Sounds simple enough. So how many *possible* lists (that is, itineraries) are there? Let's do some counting. The list contains five entries: In the first one, you can write down any of the five cities. Afterward, there are only four possibilities for the second entry (since you do not want to visit any city twice), three possible cities for the third entry, two possibilities for the fourth entry, and only one possibility (the remaining city) for the fifth and final entry on the list. In order to obtain the total number of possible lists, you have to multiply the number of first-entry possibilities (five) times the number of second-entry possibilities (four), and so on until the number of fifth-entry possibilities (one). All in all you must calculate the value of five times four times three times two times one, which gives you 120.

We now know that there are 120 possible five-city itineraries. Next, we compute the total distance for each itinerary. Suppose the five cities in question are San Francisco, Los Angeles, San Diego, Las Vegas, and Reno. In fact, let's treat that as an itinerary: from San Francisco, down to Los Angeles, continuing to San Diego, heading eastward to Las Vegas, then to Reno, and back to San Francisco. All we need to do now is take out our map of the United States and measure the total distance of this particular itinerary. Then we do the same for the other possible tours (all 119 of them), obtained by simply exchanging the positions of cities on the list. For each possibility it is quite easy to calculate the total distance. (Of course, some itineraries are absurd: for example, going from San Francisco to San Diego, and then reworking your way northward toward Los Angeles.) Armed with this knowledge of all 120 possible tours, along with their total distances, our job is practically over: We now simply pick the shortest itinerary, and *voilà*, time to start the engine.

That was quite easy. So why do computing scientists make such a big fuss over the traveling salesperson problem? Well, a five-city itinerary poses no difficulty. There are only 120 possibilities, and we can account for all of them. What happens if our enthusiastic salesperson has to make the rounds of, say, fifty cities? In this case our list has fifty entries. How many possible lists (itineraries) are there? As before, you have to multiply the number of first-entry possibilities (fifty) times the number of second-entry possibilities (forty-nine), and so on until the number of fiftieth-entry possibilities (one); the answer turns out to be 10 raised to the power of 64 (1 followed by 64 zeros). To get an idea of how "absurd" this number is, consider the following: If we were in possession of a computer able to examine one trillion possible itineraries per second (rest assured, such a computer does not yet exist) and if this computer were to have been working faultlessly since the big bang (roughly 15 billion years ago, according to current estimates), then we would have examined by now 0.000000000000000000000000001 percent of the total number of possible itineraries.

Clearly, brute force won't work. As things stand now, there are astute ways of finding rather *good* answers, but there is no known way to obtain the *best* answer. That is, we know how to find, within a reasonable amount of time (no billion-year wait), fifty-city itineraries that are short—but not the shortest. The traveling salesperson problem has a special status in computing science. It is a member of a large family that currently has hundreds of important problems—from designing efficient airline timetables to constructing good computer chips. All the members of this happy family share one thing in common. If you can solve *any* one of them in a reasonable amount of time, then you can solve *all* of them efficiently; conversely, if even one of them turns out to be really hard (that is, admitting no quick solution), then all of them are hard. By "really" hard (as opposed to merely "hard"), I mean being in possession of a mathematical proof (rather then just hands-on experience) to the effect that the problems in question require huge amounts of time to solve; on the contrary, a "really easy" verdict would follow immediately from the finding of a (mathematically provable) fast solution to *any* member of this family. Despite the continued efforts of the best minds in computing science over the past three decades, we still don't know which way the balance will turn, though our bets are placed on the "really hard" side.

This family of problems is known in computing science as "NP-complete," where the "complete" part signifies the above interdependence among all its members: the we-all-stand-or-fall-together property. The "NP" part is also quite interesting: It is a mathematical term that basically boils down to the difference between *finding* a solution and *checking* one. If you are given a fifty-city problem and asked if there exists an itinerary of less than 500 kilometers, then that's hard. On the other hand, if someone hands you a candidate solution, claiming that it's a perfectly valid tour of less than 500 kilometers, just checking this claim is a cinch (all you have to do is verify that all routes chosen do indeed exist and then compute the total distance). Evolutionary computation also exhibits the same dichotomy: *Finding* a solution is hard (that's why we use evolution), whereas *measuring* the adequacy of a *given* solution is far easier (we called this *fitness* in the context of evolutionary computation).

Suppose you could examine not one itinerary at a time, but rather several of them in parallel. An even bolder proposal was put forward in 1994: Suppose you could examine *a whole lot* of itineraries in parallel. How? It's time to intertwine the two yarns we've been weaving here, combining computing with biology: Solving the traveling salesperson problem using (real) DNA molecules. To make life easier, we shall simplify the problem by assuming that the distances between any two cities are equal. All we want to know is whether there exists an itinerary such that each city is visited exactly once, with no hightailing. (Such a tour may be easy to come by when considering developed countries that are crisscrossed with interconnecting roads, though it is far from obvious for less-developed countries.) Though seemingly easier, this problem is in fact no less hard than the original one; it belongs to the same NP-complete family.

The basic idea is to use oligonucleotides—those short chains of up to twenty nucleotides—to represent cities and roads. For simplicity, let's consider chains of just four letters (nucleotides) instead of twenty. First, we assign a unique oligonucleotide tag to each city; for example, in our five-city problem we might decide to assign the tag ATCG to San Francisco, GCCT to Los Angeles, TGCA to San Diego, CGAG to Las Vegas, and CCGG to Reno. The cities are connected by a network of roads, each of which also gets a tag. This tag is composed of the second half of the departure city's DNA tag affixed to the first half of the destination city's tag. For example, the road from San Francisco to Los Angeles will be tagged CGGC (San Francisco "donated" the CG part and Los Angeles chipped in the GC part), and the road from Los Angeles to San Diego will be tagged CTTG. Notice that the reverse route, say, from Los Angeles to San Francisco has a different tag; our setup is not symmetric. This is quite advantageous since it echoes the situation with real roads (for example, one direction might be closed due to construction work).

One last bit of molecular biology in order to complete our system: The four nucleotide letters (A, T, C, and G) have a tendency to pair up: Letter A "likes" to bind with T (and vice versa), whereas C is enamored of G (and, luckily, vice versa). In molecular biology these are known as complements: A is complementary to T, and C is complementary to G. So far, we've constructed DNA tags for the cities and roads. In fact, rather than employ the city tags, we'll make use of the city complements: San Francisco (TAGC), Los Angeles (CGGA), San Diego (ACGT), Las Vegas (GCTC), and Reno (GGCC).

Before moving on, let's take stock of our oligonucleotide inventory:

San Francisco: ATCG, and its complement: TAGC
Los Angeles: GCCT, and its complement: CGGA
San Diego: TGCA, and its complement: ACGT
Las Vegas: CGAG, and its complement: GCTC
Reno: CCGG, and its complement: GGCC
San Francisco–Los Angeles highway: CGGC
Los Angeles–San Diego highway: CTTG
(The other road tags aren't shown here.)

We next pour into a test tube a huge number of copies of the city complements and roads (or, rather, of their respective DNA tags), thus priming our computer-in-a-test-tube. What happens now? These DNA fragments behave the way they always do, following the rule of "opposites attract." The tiny DNA slivers start to bind to each other, forming longer and longer chains. Each chain represents a possible tour. The beauty of it is that only legal tours, involving extant roads, can form.

Let's see how this actually works in our example. Suppose that one of the San Francisco–Los Angeles highway tags

(CGGC) meets a Los Angeles city complement tag (CGGA). Our system is designed such that the second half of the road (GC) is exactly complementary to the first half of the Los Angeles city complement (CG); the two will thus bind at these two sites. You can picture this as a two-layer arrangement: the highway CGGC on top, with the city complement CGGA underneath—shifted two positions to the right, just below the highway's GC part.

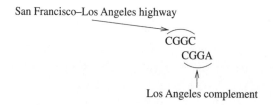

San Francisco–Los Angeles highway

CGGC
CGGA

Los Angeles complement

Now a merry CTTG tag—a Los Angeles–San Diego road tag—comes along. Upon meeting our two-layer couple, it finds two of its beloved nucleotides: The second half of the lower-level Los Angeles complement—GA—is exactly complementary to the Los Angeles–San Diego first half—CT. Thus, this tag will now bind to the couple (at the upper level), forming a triplet: The two road tags—San Francisco–Los Angeles and Los Angeles–San Diego—are juxtaposed on the upper level, with the Los Angeles complement situated at the lower level exactly in the middle.

San Francisco–Los Angeles highway   Los Angeles–San Diego highway

CGGC CTTG
CGGA

Los Angeles complement

Now you see what the city complements are for: They act as "glue" for the roads. The DNA tags are designed so that two roads can hook up with each other only if the right glue piece exists; and this is present only if the two roads *can* indeed be connected. The San Francisco–Los Angeles and the Los Angeles–San Diego roads can be glued together, whereas the San Francisco–Los Angeles and the San Diego–Las Vegas roads cannot (since they don't "share" a common city among them, and thus there will be no glue tag—no city complement—to allow them to hook up).

What we have described above amounts to the first step of the DNA computing process, aimed at solving the traveling salesperson problem for a given map (a given setup of cities and roads). We pour into a test tube copies of the tags—city complements and roads—and let Nature take its course. The DNA fragments will swim within their allotted aquarium, randomly brushing up against one another, and numerous bonds will form in accordance with the bonding rules. These bonds are in fact small DNA molecules that represent itineraries. The crux of the matter lies in how many tag copies there are. We're talking *massive* parallelism here: trillions and trillions of tags swimming in the test tube. This is what enables tours to emerge out of these random interactions.

Our job is not quite over, though. We're now in possession of numerous DNA molecules—representing various possible itineraries—many of which (if not most) do not represent viable solutions. Some tours contain too many cities, some contain too few, and still others—while containing exactly five cities—include duplicates, meaning cities that appear more than one time (remember that we wish to visit each city but once). We next subject our test tube to two molecular "sieves" that are intended to rid us of unwanted molecules—those that represent "bad" tours. First, we chuck all itineraries that do not have *exactly* five cities (no more, no less). Then, we discard all tours that do not include each of the five cities. If any valid molecules remain, then the answer to our question is yes: There exists an itinerary such that each city is visited exactly once (remember that we're talking about the simplified problem where all distances are considered equal). Otherwise, the answer is no. Filtering out the "bad" molecules is done by well-known techniques, routinely used by molecular biologists (these go by such resounding names as "gel electrophoresis," "affinity purification," and "sequencing"). The end result is a test tube that contains a (molecular) solution to our problem.

Leonard Adleman, who first carried out this experiment, thus launching the field of DNA computing, wrote: "For me, it is enough just to know that computation with DNA is possible. In the past half-century, biology and computer science have blossomed, and there can be little doubt that they will be central to our scientific and economic progress in the new millennium. But biology and computer science—life and computation—are related. I am confident that at their interface great discoveries await those who seek them."

This is, however, still a brute-force method; we're simply counting on the huge number of interactions taking place in parallel. While this may work for a small number of cities (Adleman's original experiment addressed a seven-city problem), it has been estimated that upping the number to a few dozen towns would require a test tube the size of the Earth. Clearly, even with such a vast amount of parallelism, one must still design more efficient methods of computing.

Another problem is created by errors. In our example we assumed that all operations were perfect and that everything took place exactly as planned. Alas, molecular reality is error-prone—from time to time incompatible roads may, for example, bind to each other. The San Diego–Las Vegas highway, tagged CACG, might bind to the Los Angeles complement because of a mismatched pair: A binds erroneously to A.

San Francisco–Los Angeles highway   San Diego–Las Vegas highway

CGGC CACG
CGGA

Los Angeles complement

You now find yourself in possession of a tour that includes the San Francisco–Los Angeles road immediately followed by the San Diego–Las Vegas road, a route which—barring teleportation—is impossible.

You can reduce such errors by judiciously selecting good city and road tags. This is known as the *encoding problem*: how to encode your problem using DNA molecules such that as few errors as possible occur. Max Garzon, Russell Deaton, and their colleagues from the Molecular Computing Group at the University of Memphis looked into the application of evolutionary computation in a test tube for solving the encoding problem. They used the naturally occurring biochemical reactions in the tube as a fitness measure: Encodings giving rise to erroneous bonds were considered less fit. The Memphis team thus co-opted the errors to work *for* them, by providing the selection pressure needed to drive the evolutionary process. To err may be human, but not if you're a DNA molecule under their scheme: Evolution does not forgive.

Finally, let's not forget that we're talking about "messy computing," with biological procedures that have to be carried out in the lab and a process that may require several days (the original seven-city problem required seven days of lab work). This is perhaps somewhat of a minor problem since we shall probably be able to automate the process to a large extent.

To the delight of researchers working in the field, several obstacles still remain before test tubes find their way into the computer programmer's arsenal. As Max Garzon once said: "We're currently at the ENIAC stage of DNA computing" (ENIAC, the Electronic Numerical Integrator and Computer, was one of the first computers, a room-sized beast built in the 1940s). Nonetheless, this endeavor is an exciting "small is beautiful" adventure, a race to create a new kind of computer, not merely *inspired* by Nature but indeed *built* by her.

CHAPTER 10

The More the Merrier

In a 1972 paper on many-body physics entitled "More Is Different," the Nobel laureate Philip W. Anderson cited a conversation that took place in Paris in the 1920s in which F. Scott Fitzgerald remarked: "The rich are different from us," to which Ernest Hemingway replied: "Yes, they have more money."

But are the rich really that much different?

Suppose your Aunt Hillary passes away at the ripe old age of ninety-five, leaving you her entire fortune of $70,000. You'd probably have no problem putting the money to good use: paying off that irksome mortgage, finally getting down to repairing the house, and maybe there'd even be some left over for replacing your beat-up car. Then you suddenly find out that your beloved Uncle Fred has decided to move to a Buddhist monastery in Tibet and to part with his entire fortune—$2 million—which he hands over to you, in its entirety. Two million is definitely a *lot* of money, but still you have no problem in figuring out what to do with it. Indeed, a couple of months later you find yourself reclining on the porch of your new Malibu villa, relaxing from that spin you just took in your new Ferrari. Life's sure easier with a couple of million. Then one day you receive a phone call from the head lawyer of Nell Industries. It seems that the company's founder is in fact your long lost Aunt Nell! And, as with good ol' Uncle Fred, she too has decided to renounce all her worldly goods in favor of a backwoods shack in the Rockies. Guess who's the lucky one to inherit her $40 billion fortune? You've got it: You. More is different, but so much more is beyond different: It's something entirely new (40 billion one-dollar bills stacked one on top of the other, would constitute a tower 4000 kilometers high—roughly the distance between New York and Las Vegas).

Quantitative differences become—at a certain point—qualitative ones. And this holds true not only for $1 bills or human beings with some 60 trillion cells; it also holds true for computers. How so? Imagine a tiny computer that is so small by comparison to that magnificent machine majestically reposing on your desktop that it can hardly do anything at all. This midget can carry out no elaborate word processing nor run any clever games, but can perform only extremely simple operations, like adding two numbers or deciding which of three numbers is the largest. This tiny computer is quite worthless *in itself*, just like, say, a $1 bill. But what if we suddenly amassed (or inherited) a huge number of such Lilliputian computers? What if instead of one such tiny computer, we had forty billion? Now that's a whole new ball game: It's as if you went to bed with a $1 bill in your pocket and woke up with forty billion. The welter of opportunities offered by a zillion tiny computers is as astonishing as that offered by a zillion dollars.

I'd like at this juncture to christen the little beastie simply a *cell*. I don't necessarily mean a biological cell, but rather use this term in its broad sense, whereby "cell" refers to an elemental unit of an organization (as in "cellular phone sys-tem" or "terrorist cell"). When lots of these tiny com-puters, sorry, cells, flock together, we have what I dubbed *cellular computing.*

What is cellular computing? To answer this question, I'd like to walk you through a concrete example, a problem for the computer to solve, which shows just how different stan-dard computing and cellular computing are.

The Beatles are reuniting—all four of them. (Don't ask me how.) And they intend to go on stage for a single concert, which promises to be the biggest event in the history of music. Everyone will be there. And I do mean *everyone*: all six billion people on the planet. For the grand finale they announce they'll sing either "Let It Be" or "Yesterday," but not both. The audience will take a vote on the matter, and the winning song will be selected by simple majority. How is the vote to be carried out? At the end of the show, each spectator will raise either a blue flag for "Let It Be" or a red flag for "Yesterday." A computer hooked up to a camera will then scan the audience and compute whether there is a majority of blues or of reds.

How does your standard computer go about solving this majority problem? Well, all it needs to do is to *count* the number of blue flags and the number of red flags, and then to spurt out the larger of the two numbers. Sounds simple, no? It actually is—for any desktop computer. The interesting part is *why* this is so, that is, why does a standard computer solve this problem so easily.

Your run-of-the-mill computer exhibits three character-istics that are of interest to us here and that render facile the solution of the majority problem: It is *complex*, it operates in a *sequential* manner, and it has a *global* world view. By "complex" I mean that a standard computer (or—to be more precise—the so-called microprocessor hidden within its in-nards) is a much more complicated device than our simple cell. It can carry out complex computations, such as counting six billion votes, whereas, as we noted, a cell can do very little in and of itself. By "sequential" I mean that this (com-plex) computer does one thing at a time, in our case process-ing the votes in sequence, one after the other. (We came across the sequential nature of a standard computer in Chap-ter 3 when comparing its workings to the parallel operation of the brain.) By "global" I mean that a standard computer has access to all the information necessary to solve the prob-lem. It "sees" the big picture. In our example literally all six billion flags are available to the computer for the counting.

That is why any desktop computer can solve the major-ity problem quite easily: It is complex enough to count six billion votes—all of which it can "see"—by tirelessly pro-cessing them one by one. Complex, sequential, global are the three fundamental design principles of any off-the-shelf computer. This makes my job of explaining cellular comput-ing quite easy: It's just the opposite. Replace complex with simple, sequential with parallel, and global with local, and there's your equation: Simple plus parallel plus local equals cellular computing.

So how would a cellular computer solve the majority problem? First off, we need the computer itself, that is, the ensemble of simple cells. Well . . . the audience *en masse* can act as the computer, with each and every spectator being a cell! "But hey," you may be thinking, "you just said that cells are very simple devices, whereas a human being is quite a complicated creature." True, humans are complex, but we shall treat them *as if* they were simple. To be more precise, we'll restrict each and every spectator to doing only very simple things, which we'll call *rules of behavior*. In the majority problem, these rules will involve only observing neighboring flags and changing the color of one's own flag. Not only will these rules be very simple, but they will also be highly local. This means that each (human) cell will only be allowed to look at the color of her own flag and that of her four neighbors: the one to the left, the one to the right, the one up front, and the one behind. (Let's assume the audience is conveniently organized in a two-dimensional chessboard pattern.) Here's an example of such a simple rule of behavior: Observe your flag's color and those of your four neighbors: If you see more blues, then raise a blue flag; otherwise raise a red flag. You may notice that this rule is actually majority on a (very) small scale: Each spectator changes the color of her flag according to a five-hand majority—herself and the four goodly neighbors. We're almost done describing our computer. Remember that the cellular-computing "equation" involves three constituents: simple, local, parallel. Well, we've got two out of three: cells (spectators) that behave according to simple, local rules of behavior. Now suppose that they all act in unison, that is, all six billion spectators inspect their neighbors' flags and simultaneously change their own flags' colors according to the local rule. In computing parlance we call this "operating in parallel."

So there you have it: a "computer" composed of six billion simple cells, each of which is extremely myopic, with the whole lot singing in concert. Cellular computing in a nutshell—or almost. I neglected to mention one very important thing: the solution. The standard computer simply counts the votes, compares the blues count with the reds count, and then announces the winning song. How does our cellular sea-of-humans actually resolve the problem of determining the majority? As with a washing machine, we can let our computer go not through one "wash cycle" but through several such cycles: The examine-your-neighbor-and-change-your-flag step takes place not once, but several times. Through the eyes (or cameras) of an orbiting satellite this would look like a blue-and-red chessboard, with the colors of the squares in constant flurry. But will this agitated human carpet ever settle down? After all, this is a computer, and we do need a solution to the majority problem. For example, if after several cycles every single spectator were holding up a blue flag, then the Beatles would immediately see that unanimous agreement has been reached: Let it be "Let It Be." To wit, reaching a point where the human carpet is all red or all blue makes up an answer. If only things were that simple.

Before going into the problematics, let me recap the very different ways in which the standard computer and our cellular computer solve the majority problem. The way a standard computer works is actually quite easy to understand since it echoes the natural way *we* might go about solving the problem: One guy (the chief counter) counts the votes one by one. Our cellular computer, on the other hand, works in a much less natural way. There is neither chief nor boss; somehow the solution to the problem should emerge by having each cell examine its little corner of the world and raise either a red flag or a blue one. It's very easy to imagine *one* such cell, but can you imagine *six billion cells working in parallel*? Yet that's how cellular computers work!

Or don't work. As I mention above, each of the spectators performs what is essentially a local, five-hand majority computation, but we want the *global* majority over the entire six billion spectators. Alas, the local-majority rule of behavior does not lead to a solution to the global-majority problem. This demonstrates one of the prime challenges faced by designers of cellular computers: finding local rules of behavior, the operation of which—in a multitude of cells—ends up in a solution to a global problem. As a matter of fact, Mathieu Capcarrère, Marco Tomassini, and I showed in 1996 that the majority problem *can* be solved *à la* cellular computer, though the solution is somewhat trickier than the simple five-hand majority.

In recent years researchers have used adaptive methods, such as evolution and learning, to find local rules of behavior for solving global problems. In my 1997 book *Evolution of Parallel Cellular Machines: The Cellular Programming Approach*, I applied evolutionary computation to the evolution of cellular computers. In our human cellular computer, each of the six billion (human) cells changed the color of its flag according to the prescribed rules of behavior. To apply evolution, I first encoded these rules in the form of an artificial genome and then decided on how to apply the genetic operators of crossover and mutation. Then came the issue of measuring fitness: If you want to evolve a cellular majority solver, then you test your evolving cellular computer on various flag arrangements. Some of these "carpets" will have more blue flags than red ones, and some will have more red flags than blue ones. For each of these arrangements, I let the evolving cellular computer run in accordance with the rules of behavior prescribed in its genome. The more arrangements it got right, the higher the fitness, where "getting it right" meant that the cellular computer output a correct answer: "Let It Be" if there were more blue flags and "Yesterday" if there were more red flags. All the well-known ingredients of simulated evolution were there: a population of genomes, genetic operators, and fitness measurement.

Even with adaptive methods, though, getting cellular computers to work is far from easy. So why bother? The answer is because they can potentially manifest various highly desirable qualities. Cellular computing is at heart a paradigm that aims at providing new means for doing computation in a more efficient manner than other (classic) approaches. And what do I mean by "more efficient"? To a computing professional this might be one (or more) of several things: a computer that is faster or cheaper, a device that uses less electric power (this is especially important for portable devices, where battery life span is crucial), a machine that is able to store more information (that is, has a larger memory), or perhaps a computer that is able to solve hard problems—such as the traveling salesperson problem—better than any other.

We've encountered a number of examples in this book which fall under the cellular-computing umbrella. The human chessboard of changing colors may have reminded you of the configurable processor (Chapter 6). Indeed, this is an example of a cellular computer, albeit with only a few thousand cells (rather than billions). The swarm of fireflies we encountered in Chapter 6 is an example of cellular computing in Nature, with the Firefly machine being its configurable-processor analog. DNA computing is also a form of cellular computing: The DNA molecules, which represent the elemental cells in this case, interact locally in simple ways on a vastly parallel scale (billions and billions) in the test tube. You may be thinking to yourself, how dare I call a DNA molecule a *simple* cell. After all, it's quite a complicated gizmo. Well, the same way I dared to call a human spectator simple. It's not about the inherent complexity of the cells we're using; it's about what kinds of operations they're allowed (by the system designer) to carry out. The DNA-computing researcher, Richard Lipton, summed this up nicely when he wrote: "Our model of how DNA behaves is simple and idealized. It ignores many complex known effects but is an excellent first-order approximation." (As noted when discussing artificial evolution, computing scientists are fond of abstracting away "irrelevant" details. For example, the ubiquitous transistor—the basic building block of modern microprocessors—is usually regarded as a simple switch, with the complex physical phenomena taking place at the atomic and subatomic levels being immaterial.)

From swarms of fireflies to cells in your body, examples of cellular computing in Nature abound. Even as you read these lines, one such example might be meandering up your pantry: ants. Ants are social insects that live in colonies, their individual behaviors geared more toward the perpetuation of the colony than to that of an individual member. Be it reproduction, nest building, or foraging behavior, an ant colony seems to function like what entomologists call a superorganism. In particular, a colony of ants excels at finding the shortest path from their nest to a food source.

While walking from their nest to your kitchen and back, ants deposit a substance called *pheromone* on the ground, thus forming a pheromone trail. Ants can smell pheromone, and when they roam about, they tend to choose paths marked by strong pheromone concentrations. The pheromone trail allows the ant to find its way to the food source and then return with its nutritional payload. Entomologists conducted simple experiments that demonstrated that ants using pheromone trails can not only find a path to the food source, but they can find the *shortest* path.

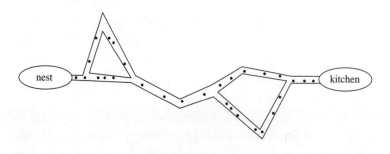

The first ants to arrive in your kitchen are those that took the two shortest branches. When they return to the nest, they will find more pheromone on the short branch and will thus not only traverse it but also increase its delicious pheromone odor. In time, most of the ants will smell their way over the shortest path.

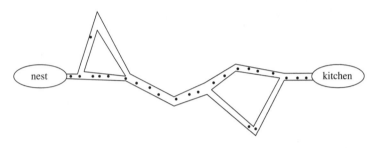

Drawing inspiration from pheromone-following ants, Marco Dorigo from the Free University of Brussels and his colleagues have come up with an approach called *ant colony optimization*. They use colonies of artificial ants to solve hard problems such as the traveling salesperson problem. The ants wander from city to city leaving an artificial pheromone trail as they do so. An ant that visits all the cities and covers less ground (literally) deposits more pheromone along its itinerary; this tour thus becomes more desirable for future ants. Dorigo and his colleagues thus used this ant-inspired approach to solve the traveling salesperson problem, as well as a host of other hard problems. But is this cellular computing? The answer is yes. The cells, in this case, are artificial ants. They are *simple* creatures that work in *parallel*, with each myopic ant able to "smell" only its *local* neighborhood; all three criteria for cellular computing—simplicity, parallelism, locality—are thus fulfilled.

Social insects such as ants display what the French zoologist Pierre-Paul Grassé called *stigmergy* (from the Greek *stigma*: sting and *ergon*: work): the accomplishing of collective tasks through indirect environmental interactions. Ants do not talk directly among themselves to spread the good news about your kitchen; they communicate indirectly by leaving a pheromone trail. Another example of stigmergy is nest construction by a swarm of wasps. An individual wasp does not have the plan of the nest in its mind but rather reacts to local structural configurations of the nest under construction. The wasps use hexagonal bricks made of wood pulp to build their comblike nest. When a wasp arrives with a new brick, it does not place it at random, but reacts to the half-built nest. The wasp is much more likely, for example, to place that brick within an existing row of bricks than to start a new row; the local brick configuration thus acts in a stigmergic manner by encouraging the wasp to fill the hole. Guy Theraulaz from the Université Paul Sabatier in Toulouse, France and Eric Bonabeau from the Santa Fe Institute in New Mexico simulated artificial, nest-building wasps in a computer, and showed that the resulting nests bore a striking resemblance to natural wasp nests. Though Theraulaz and Bonabeau were interested in modeling a natural phenomenon (nest construction in wasps) and not in building computers, such swarms of simple artificial wasps working locally might one day give rise to yet another form of cellular computing.

"What about artificial neural networks?" you may be asking at this point. "Aren't they based on the workings of a host of small, tiny processors—which we called artificial neurons?" Actually, artificial neural networks do not fit the bill. They exhibit only two elements of our three-pronged cellular-computing equation: Yes, the basic cells (artificial neurons) are simple, and, yes, the neurons interact in a highly parallel fashion, but, no, these interactions are not necessarily local. Just by looking at the images in Chapter 3, we can see that artificial neural networks are highly interconnected, with each neuron talking to many neighbors, whereas we insisted that the neighborhood be kept cozy where cellular computing is concerned. Thus artificial neural networks are generally not cellular computers.

Keeping the neighborhood of a cell small, so-called local connectivity in computing jargon, is one of the greatest boons offered by cellular computing. As computing scientist Danny Hillis wrote back in 1985: "As switching components become smaller and less expensive, we begin to notice that most of our costs are in wires, most of our space is filled with wires, and most of our time is spent transmitting from one end of the wire to the other.... Most of the wires must be short. There is no room for anything else." And short wires are exactly what is required by local connectivity.

Artificial neural networks can, in fact, go cellular, as shown by Leon Chua from the University of California, Berkeley and his colleagues, who came up with the idea of *cellular neural networks* in 1988. These are artificial neural networks connected in a cellular fashion: Each neuron is placed in one square of a two-dimensional chessboard and can communicate only with its direct neighbors on the board. The cellular neural network thus fits the cellular-computing bill. The artificial neurons are simple, they work in parallel, *and* they are locally connected.

Chua and his colleagues designed a cellular neural-network chip, which finds uses in domains such as *image processing*. We humans rely greatly on our visual sense (large portions of our sensory cortices are devoted to processing visual information), and thus imbuing computers with the ability to handle images is of the utmost importance. "Handling images" includes a wide range of tasks, from recognizing faces in a picture to identifying straight lines and circles. Tasks such as face recognition are known as "high level," while tasks such as shape identification are known as "low level"; image processing deals mainly with low-level tasks. With a cellular neural network you can, for example, do so-called contour extraction.

The contours of the left-hand image of the most famous smiling woman (Figure 10-1) are shown in Figure 10-2. Such low-level operations are used by the computer when it attempts to perform higher-level tasks such as face recognition. Suppose you show me a picture of a woman standing in front of a shack, with the Rocky Mountains in the background. "Who is this woman?" I ask, and you answer immediately: "Why, it's my Aunt Nell, of course." This mundane exchange, which happens whenever you show your photo album to a friend, becomes arduous when a computer tries to do the same. The computer first scans the photo using a camera or a scanner; that's the easy part. Before it can proceed to

Figure 10-1   Mona Lisa

Figure 10-2   Contours
of left-hand image of
Mona Lisa

177

identify the woman in the picture, however, it must find out *where exactly she is located in the image.* Distinguishing between a face, a shack, and a mountain is trivial for us, since we are totally oblivious to the heavy-duty processing taking place in our brain during the half-second or so it takes us to recognize a face. With a computer, though, we are quite aware of all this intense processing since we have to program it explicitly to be able to recognize Aunt Nell. And to do so, we first apply various low-level operations, such as contour extraction: That's how the computer can greatly reduce the amount of information it needs to imbibe by concentrating on the "small oval shape with two symmetrically opposed holes," which it can identify according to its contours. Having analyzed the image with low-level operations and finding out the region of interest in the photo, the computer can finally begin trying to identify the person itself. (Face recognition and image processing both benefit these days from adaptive methods, such as learning and evolution. But then, I guess you suspected that by now.)

If cellular computers are easier to build (because cells are simple and connections are local) and if they offer potentially wonderful gains, then where can you buy one? Alas, you cannot (yet) walk into your corner computer shop and buy a cellular computer with billions and billions of cells. But, echoing Arthur James Balfour's words concerning Winston Churchill—*"I thought he was a young man of promise, but it appears he is a young man of promises"*—I firmly believe cellular computing is a young field of promises. There is already a legion of evidence out there, though for the moment mostly in the form of research papers and projects. Of course, it is always encouraging to consider the success of the ultimate manufacturer of cellular computers—Nature.

CHAPTER 11

Biologic or "By Ole Logic"

An eagle flaps its wings; a Boeing 747 doesn't. A dolphin wiggles its body and jiggles its fins; a submarine just has a motor in the back. A dog walks on legs; a Mercedes-Benz rolls on wheels. A rose runs on water and light; a flashlight runs on batteries. A tiger develops in a womb from a single cell to a magnificent multicellular beast; a toy tiger is constructed full blown in a factory. A piano player goes through years of intensive training, learning to hone her talent; a piano learns nothing. *Homo sapiens* have evolved by means of natural selection; watches are designed by watchmakers.

Engineers and Nature have usually taken distinct routes in their creation of complex objects, differing both in the final artifacts produced as well as in the design process itself. And the recent movement that seeks inspiration in Nature has come up not only with novel objects but also with entirely new ways of designing objects. Thus, current-day robots may possess legs, fins, or wings; electronic circuits may develop in a manner akin to that of multicellular living beings; watches can heal themselves; computers can learn to play a mean game of backgammon; and bridges can be evolved.

Having visited several lands in the *Terra Nova* of computing and having acquired along the way many new colorful approaches, we shall now use these colors in the remainder of the book to paint the big picture. In this chapter I'd like to take a closer look at the main differences between human's work and that of Nature, specifically focusing on how these differences relate to our current engineering efforts. When does it pay to be biological, and when is it better to use the traditional, by-ole-logic way? As a concrete example I'll consider two different kinds of flying machines: birds and airplanes. When engineers set about to design an airplane, they proceed in what is known as a *top-down* approach: They start with the general issues and questions (the top) and go all the way down to the nitty-gritty. At the top there is the decision—usually made by senior management—to build a new airplane. Next comes the *requirements analysis* phase, which basically answers the question: *What* should this new machine be able to do? It might be required, among other things, to carry up to 600 passengers, to take off and land on short runways, and to handle severe weather. Having defined the *problem*, it is time for the engineers to enter in force, their job being to find a *solution*; now that we know *what* we want, it's time to see *how* we go about building it.

The design process continues in a top-down fashion, breaking the big problem down into smaller and smaller subproblems; one doesn't jump immediately to the nuts-and-bolts level. This breaking-down process might be done by identifying key parts—such as the cockpit, the fuselage, the engines, and the wings—and assigning their design to different teams (which obviously must cooperate among themselves; after all, there is but a single final object being built: the airplane). Each such key design problem is further divided into smaller subproblems; the wings team, for example, will be considering flaps, spoilers, ailerons, and other such beasties. The design process is by no means simply a forward march; often one must go back to the drawing board since the part in question doesn't function as it should. This back-and-forth process ends up with a design specification—a complete plan of the airplane (such a complex object might require years of design work). Now it's time to fabricate the machine, a task which in itself may be quite elaborate for such an artifact. It might, in fact, require a separate design process since in all likelihood new fabrication techniques for the new airplane will have to be developed.

Engineering designers thus start out with a clear top-level goal in mind, and then work their way downward toward the most minute details, ultimately coming up with a comprehensive solution. Nature works quite differently. For one thing, Nature has no explicit, a priori goal; Nature does not embark upon a lengthy R & D project whose final objective is the construction of a bird. Nature employs evolution, and evolution is shortsighted: The only goal, the only thing that matters, is immediate survival. Nature, if any designer at all, is a blind one at that. The ability to fly emerges over eons since it confers some advantage to the animals that possess it. Thus, when speaking of evolution's goal, one can at best describe it as an implicit, short-term one: survival. (In *The Blind Watchmaker* Richard Dawkins proposed a way by which wings might have evolved. His scenario starts out with wingless animals that leap between tree boughs. Small flaps of skin that help extend the jump or break the fall—by acting as an airfoil—will bestow an immediate survival benefit upon

their owner. Little by little, over the course of many generations, the accumulation of small, ever-better modifications to
these flaps might end up as full-fledged wings.)

Evolution is further distinguished from engineering in
that it is a *bottom-up* process: Its "products" emerge from the
myriad of interactions that take place in the biosphere. There
is no top-down process that starts out with a major, farsighted goal that is then broken down successively into
smaller and smaller subgoals, until they become doable. There
are just numerous interactions, both among organisms and
between them and the elements, out of which emerge all the
wonderful devices we see around us (and in us), such as
wings, eyes, feet, nervous systems, and rock stars.

Nature's open-ended, shortsighted, bottom-up style as
opposed to engineering's guided, farsighted, top-down approach is the crux of the difference between the two. It entails
several other distinctions between the engineering enterprise
and Nature's workings.

Engineers usually seek not only to create a widget that
works, such as an airplane or a coffee machine, but indeed
one that works well; often they evoke terms such as "efficient"
and "optimal" to describe their desired product. Nature, on
the other hand, cares nothing for these qualities; designs need
not be the best or the fastest or the most efficient; rather,
Nature is after "just-do-the-trick" solutions, namely, ones
that can survive. If an organism has even the slightest advantage over its confreres, then that's all it takes; it'll be the winner in the survival race, and its genes will pass on to the next
generation.

"But how then," you might be asking yourself, "has Nature come up with all those marvelous designs we see out
there—such things as seeing gadgets, delicate manipulators,
and thinking machines, which are still way beyond our cur-

rent engineering capabilities?" First off, let's not forget that Nature has had a bit of a head start—3.5 billion years to be precise. This figure should not be brushed aside lightly: It is a *huge* amount of time, practically impossible for us to grasp. As noted by Charles Darwin in the *Origin of Species*: "The mind cannot possibly grasp the full meaning of the term of a hundred million years; it cannot add up and perceive the full effects of many slight variations, accumulated during an almost infinite number of generations." Our inability to grasp such a vast period of time is not so surprising if you think about the environment in which our minds have evolved to function. During most of our evolutionary history, there was no survival value in being able to comprehend the expanse of a million years (nor, for that matter, of a millionth of a second). It is only very recently (no more than a few thousand years) that we have begun dealing with such huge numbers, our minds coming to appreciate time out of mind. While for engineers time is of the essence, for Nature the essence is time.

In coming up with her flying machine, Nature thus spent a little more than the few years engineers spend in designing a Boeing 747. The chirping critters we see today outside the window are superb beasts, yet their beginnings—the ancestral forms that flew the Earth millions and millions of years ago—were probably much less impressive. It's hard to match our current engineering achievements with those of Nature, but then again, it might also be somewhat unfair. We should probably compare our current-day devices not with modern flora and fauna, but rather with Nature's first attempts, those that had been in existence so many millions of years ago (and which are now—for the most part—extinct).

Nature not only takes her time but also makes use of a huge amount of resources. Charles Darwin remarked that the evolutionary process goes on "for millions on millions of years; and during each year on millions of individuals of many kinds...." While an engineer usually tries to cut costs wherever possible, Nature is lavishly wasteful. She works by trial and error, indeed *lots* of trials and *lots* of errors. Charles Darwin quoted Milne Edwards as quipping that "nature is prodigal in variety, but niggard in innovation." There are many more extinct species than surviving ones, or, as Richard Dawkins said: "However many ways there may be of being alive, it is certain that there are vastly more ways of being dead...."

Evolution is basically a forward process: Any new entity must be immediately functional, or else it dies out. As we've seen above, engineers can (and often do) go back to the drawing board in order to fix a flawed design. Nature, on the other hand, cannot move backward; there is no drawing board to go back to, no possibility of deciding, "Well, this new wing design isn't so good, so let's go back to the old one and try to improve it in another way." In Nature, no good means no life (as in dead).

Another difference between engineered devices and natural ones has to do with "leftovers." In human-made systems essentially every single part is accounted for and serves some purpose; if not, then it is removed without further ado. Nature, on the other hand, tends to accumulate junk, her motto being: "If it's not harmful, then it's none of my business." Why waste effort on removing innocuous parts? Modern creatures thus carry vestiges of past epochs, which might have served some purpose at one time, but which are totally useless today (our tail bones, for example).

Let's take stock of what we've gleaned so far about the biological versus the by-ole-logic. When engineers design a product, they have a clear goal in mind; they proceed in a top-down manner, seeking to create an artifact that is—as much as possible—the best solution to the problem at hand. Nature, on the other hand, has but a single, short-term goal in mind, survival; she relies on the process of evolution to "design" her products, slowly proceeding in a bottom-up manner, sparing no expense and taking no heed of her extravagant wastefulness. With respect to expenditure one might say that engineers are like Ebenezer Scrooge whereas Nature is like Santa Claus. In a nutshell, Nature designs by evolution, while engineers design, well, by design.

Nature has come up not only with ingenious solutions to specific problems—for example, structural designs such as eyes or wings—but indeed has found (and founded) entirely new *processes* to aid in the emergence of complex organisms. Two of the most important ones are ontogeny (the development of a multicellular organism from a single mother cell) and learning.

Engineers and computing scientists have been turning of late more and more toward Nature, wishing to learn from her ways and means. In building novel artifacts, they seek inspiration in a wide range of phenomena, from general processes, such as evolution, ontogeny, and learning, to more specific natural inventions, such as immune systems, eyes, and ears.

Why are we so enthralled by the biological? After all, the by-ole-logic way is methodical and precise, while the biological is so much "mushier." Think of (or in my case imagine) that sleek, black Porsche 911, comfortably reposing in your garage—a triumph of modern engineering. Since every step of its design and construction involved traditional engineering techniques, we know exactly what it is capable of, and of what it is incapable: how fast it can go, its fuel efficiency, its ability to withstand shocks, its maneuverability along curves, its braking distance, and so on. Contrast this with Nature's creations, where we are often at a loss to answer such questions as: Does it work; if so, why? If not, why not? Does it work *well*? Does it work well all the time? How far can we push the system? What are its limits? We know how to answer such questions when it comes to a Porsche, whereas a dung beetle presents us with a far more difficult case.

You could argue that a dung beetle is a problem for biologists, whereas we're interested in a "hard" engineering problem, building Porsches. The problem is that once we move from the by-ole-logic to the biological, using techniques such as those described in this book, we find ourselves on murkier grounds. Consider the robots discussed in Chapter 4, whose brains consist of artificial neural networks that emerge by means of evolution. We find ourselves faced with an engineered machine—the robot—for which we are very hard put to answer all those questions of the previous paragraph (we elaborate on this issue when we talk about scigineering).

It might seem that I come to bury the biological, not to praise it: Why use those mushy, biologically inspired techniques to build Porsches when we have such good, well-known classical methodologies? Well, despite appearances to the contrary, most of our engineering achievements to date are quite simple, at least in comparison to Nature's. A Porsche is less complicated by far than a dung beetle; in fact, I'd probably be risking very little in claiming that a Porsche is simpler than any one cell of your body! Our engineering techniques have worked wonders in erecting modern civilization, but our appetites keep growing; technology feeds upon itself by creating new niches that bring about new needs and desires for more technology.

The more elaborate our artifacts become, the more difficult it is to find solutions by using only traditional computing and engineering techniques. That's when we supplement the by-ole-logic with the biological. Notice my use of the term *supplement*: We're not rushing to chuck the ole techniques; rather, we want to eat the cake and have it too, combining the by-ole-logic and the biological. There's no point in being a traditionalist or a Young Turk just for its own sake; the goal is to build better artifacts, whatever the means.

And just what good is the biological to engineers? We've been answering this question throughout most of this book; let's try to summarize some of the benefits we've encountered. As I've just remarked, technology keeps getting more and more complex, which means that our traditional methodologies run up against a wall much sooner than they did before; more and more often they are overstretched to their limit—and then some. That's when we start considering the biological, which often permits us to make do with but a partial design—to be completed through evolution, learning, and other biologically inspired techniques. (Incidentally, even automobile companies have recently started employing techniques such as evolutionary computation and artificial neural networks to design certain parts of their cars.)

When the by-ole-logic is stretched to the limit, it's worth trying the biological, though one must remember that it is not a panacea. I hope I've managed to convey the intricacy of applying these techniques in the preceding chapters. It's not easy to get a good bridge to evolve or to have a robot learn to walk.

Another salient difference between Nature's devices and those of humans has to do with their robustness. This term means different things in different domains, but it basically boils down to the ability to cope with a wide range of circumstances. Place a cockroach in virtually any imaginable terrain, and it'll have no problem in walking the Earth; a robot, on the other hand, has a much harder time breaking new ground. (As we saw in Chapter 4, the robotic soccer teams played much better at their home institutes than at the match site, having grown accustomed to the home terrain.) You can suffer a severe blow and still keep on ticking; the same cannot be said of your Porsche. Plants have an uncanny ability to grow toward the light, wherever it may be. A computer recognition system has a much harder time than a human in identifying a previously bearded man who suddenly shows up clean-shaven. From bacteria to brains, there are endless examples of just how robust natural creatures are, a quality that we'd like to instill in our artifacts.

Nature places its creatures in a continual lifetime struggle for survival. Moreover, every living creature today comes from a long line of distinguished ancestors that had one thing in common: They were survivors (at least long enough to engender a dynasty). Small wonder they've evolved to be so versatile. After all, robustness is decidedly a boon to survival.

To emphasize just what it means to pass through the evolutionary sieve, let me recount a short tale. The 11 o'clock news announces the founding of a new airline company whose rates are three times cheaper than the cheapest of airlines. How does it manage? Simple: no humans! At Robo Airways every job—onboard personnel, reservation clerks, ground crews—is handled by computers and robots. Would you fly the robotic skies? I'd bet the company would go bankrupt very quickly for one major reason: No one would want to fly without a human pilot aboard. Why is that? After all, any modern-day aircraft has an automatic, onboard pilot that performs much of the drudgery of piloting, and you don't have to stretch your imagination too far to envisage a fully automated flight system. What's so special about a human pilot? Well, it's not so much the piloting abilities as the pilot's humanness. Obviously, there is a psychological angle that comes into play, a human pilot being much more similar to us than a machine. Let's dig a little deeper, though.

According to robotics researcher Rodney A. Brooks, an examination of the evolution of life on Earth reveals that most of the time was spent developing basic intelligence. He wrote that: "This suggests that problem solving behavior, language, expert knowledge and application, and reason are all rather simple once the essence of being and reacting are available. That essence is the ability to move around in a dynamic environment, sensing the surroundings to a degree sufficient to achieve the necessary maintenance of life and reproduction. This part of intelligence is where evolution has concentrated its time—it is much harder." Playing chess, reading newspapers, and piloting airplanes are very recent skills that piggyback on our versatile brains, which have evolved over millions and millions of years. The title of Brooks's paper—"Elephants Don't Play Chess"—nicely captures this idea: While not able to play chess, elephants are nonetheless robust and intelligent, and they are able to survive and reproduce in a complex, dynamic environment.

When Nature comes up with a new product line, it is immediately subjected to the most grueling series of tests ever invented: evolution. That's why we can trust the human pilot much more than we can the automatic one: Piloting skills are but a mere add-on to a powerful system whose design has been millions of years in the making. Or, consider another example: Any human can tell the difference between a baby and a doll, our visual system having evolved to be able to keenly distinguish our kin. Yet with Dean, the housemaid robot of Chapter 4, this is far from obvious. How can we be sure it won't confuse one with the other (with the consequences being anything from comic to disastrous)?

The biological approach to engineering is a powerful sword to be wielded when the old tools fail, or when they yield unsatisfactory solutions. Applying processes such as evolution and learning does have its price, though, since we've seen how lavish the biological tends to be. We do have, however, the benefit of very fast artifacts, such as computers; thus, the biological, when applied to engineering, need not necessarily take millions of years (as with natural evolution) or years (as with human learning). Moreover, the biological approach has the potential of yielding more robust solutions, ones that do not fold with the slightest breeze. And let's not forget that another possible biological approach to engineering is to seek inspiration not in Nature's grand processes but rather mimic some of her solutions, examples of which are artificial retinas and artificial cochleae.

As I've remarked above, we need not replace the by-ole-logic with the biological but rather combine the two, thus enjoying the best of all possible worlds. And when opting for the biological, we don't necessarily have to remain 100 percent faithful to Nature; we can even at times take a bio-*illogic* path. Let me give just one example, that of Darwinian versus Lamarckian evolution.

The Chevalier de Lamarck was an eighteenth-century intellectual who argued in favor of evolution many years before Darwin. In this he was right. What he got wrong was the mechanism, now known as Lamarckism, or Lamarckian evolution, which is based on two principles: the principle of use and disuse and the inheritance of acquired characteristics. The first principle asserts that those parts of an organism's body that are used grow larger, and those that are not used tend to wither away. The second principle states that such acquired characteristics are then inherited by future generations. Thus, a bodybuilder bequeaths his developed muscular physique to his children. Or, consider the following story about giraffes: The early ones had rather short necks, and so they strained desperately to reach leaves high up in trees. These mighty efforts resulted in longer neck muscles and bones, which they passed on to their offspring; each generation of giraffes thus stretched its neck a bit, a head start which it passed on to its offspring.

Lamarckian evolution seems reasonable. In fact, it seems rather enticing: Wouldn't it be great to have—from day one—all those acquired characteristics of your ancestors? Alas, that's not how things work, and so the Darwinian theory of evolution has supplanted the Lamarckian theory. The giraffe does not directly pass along its long neck—acquired during its lifetime—to its offspring. Darwinism is more roundabout: Some giraffes are genetically predisposed to develop into mature animals with long necks. These will then have an advantage (however slight) over others, since they will be able to reach higher leaves. Thus, they will stand a better chance of surviving and leaving offspring, which will in turn inherit the genetic predisposition (which might then be further enhanced through favorable mutations).

While the biological theory of evolution has shifted from Lamarckism to Darwinism, *this does not preclude the use of Lamarckian evolution in artificial settings.* It can greatly accelerate evolution since a good acquired trait can be immediately incorporated into the genome. There is still a debate as to the use and usefulness of artificial Lamarckian evolution, though my intention here has simply been to show that we need not remain 100 percent faithful to Nature.

The biological blazes new trails that lead to fascinating lands. But the lesson to take home is that whether by-ole-logic, bio-logic, or bio-illogic, what matters is the end result: By hook or by crook, just get it to work.

CHAPTER 12

Who's the Boss?

Tyger! Tyger! burning bright
In the forests of the night,
What immortal hand or eye
Could frame thy fearful symmetry?

Who indeed framed the tyger? Two hundred years after William Blake wrote the beautiful opening stanza of *The Tyger*, we have a better idea of how tigers come about: through the process of evolution. There is no a master plan or a "hand of god" or any ultimate goal; the driving force is the short-term objective of survival, with the process consisting of the slow accumulation over millennia of numerous small—yet profitable—variations. In the forests of evolution burn many a creature, with Nature's immortal evolutionary hand slowly framing the fearful symmetry of the tiger.

Natural evolution is an open-ended process and is thus distinguished from artificial evolution, which is guided, admitting a "hand of god": the (human) user who defines the problem to be solved. When we apply evolution with a well-defined goal in mind—such as designing a bridge, construct-ing a robotic brain, or developing a computer program—what we are doing is akin to animal husbandry. Farmers have been using the power of evolution for hundreds of years, in effect doing evolutionary computation on domestic animals. In order to "design," say, a faster horse, they mate swift stal-lions with speedy mares, seeking to see even faster offspring emerge from this coupling. This is quite similar to the use of evolutionary techniques discussed in this book: The farmer defines the fitness criterion (say, speed) and performs the selection process by hand (by choosing the fastest individ-uals in the equine population); he then lets Nature work out the genetic details involved in the coupling act. It's rather interesting to note that farmers and breeders had started using this method long before either evolution or genetics came under the scrutiny of science.

Farmers can start out with slow horses and evolve fast ones, but can they evolve tigers? Engineers can start out with bad bridges and evolve good ones, but can they evolve a town like Cambridge? Robotics researchers can evolve robots that manage to amble decently, but can they evolve a robotic housemaid? Programmers can evolve computer programs that solve various well-defined problems, but can they evolve truly intelligent software? Natural evolution has done it all: Complex organisms, sophisticated structures, intelligent beings; it did so by being open-minded, ready to accommo-date any improvement that came along.

The Merriam-Webster online dictionary (www.m-w.com) defines "open-ended" as something that is "not rigorously fixed: as **a:** adaptable to the developing needs of a situation **b:** per-mitting or designed to permit spontaneous and unguided responses." Open-endedness is thus the flip side of guided-ness, and it is a crucial aspect of natural evolution. Nature, having no specific goal in mind, can easily change course so as to face the winds of change, and in so doing explore numerous designs out of what is essentially an infinitude of possibilities. Humans, on the other hand, even when using evolutionary techniques, *do* have an ultimate goal in mind, be it a retractable bridge or a program that computes taxes.

When we apply evolutionary techniques, the ingredients are all there: a (possibly huge) population of individuals, survival of the fittest, and the equivalent of genetic operators. Yet the hand of god is ever present in the background. At every step of the way an individual's fate is decided in accordance with its ability to perform in the arena set up by the puppet master, and the master wants his puppets to do some very specific tricks. This places a fundamental a priori limit on what evolution can achieve: If we set about to find fast horses, then we might succeed in doing so, but we'll not suddenly see the emergence of tigers.

Nature's open-endedness runs deeper, though, than the mere absence of a goal and a god—of a teleology and a master. It has to do with Nature's ability not only to play the game but indeed to change the rules altogether. Let me drive this point home by way of a sports example. The game of basketball is played on a court 90 feet long by 50 feet wide between two opposing teams of five players, who score by tossing an inflated ball through a raised goal. The rules are well known and rigid, with changes being rare, minor (for example, adding the three-point shot), and human-mediated (say, the NBA committee). This scenario is analogous to that of guided evolution: A human designer sets the stage (or in this case court) that gives rise to a (fiercely competitive) evolutionary process, from which but one kind of creature may emerge: basketball players. The process is not open-ended since there is a precisely defined goal (scoring more points), with virtually immutable rules. Though superb basketball players can (and do) evolve, this arena does not give rise to first-rate opera singers.

Playing Nature's "basketball" game is quite different. For one thing, there is no clear objective; at best, one can speak of a very basic goal, that of coming out of the match alive. What's more—and this is where open-endedness comes into play—Nature keeps changing the rules of the game, both in time and in space. Being 7 feet tall might be good at a certain place and time, whereas elsewhere or at another time it might be downright deleterious. And sometimes the rules are such that having a superb tenor voice is a match winner. Nature's game of basketball is more of a metagame, where you want to score more points, but have to figure out how points are scored.

In Chapter 1 we discuss an important distinction in Nature, that between genotype and phenotype. An organism's *genotype* is its genetic constitution, the DNA chain that contains the instructions necessary for the making of the individual. The phenotype is the mature organism that emerges through execution of the instructions written in the genotype. It is the phenotype that engages in the battle for survival, whereas it is the genotype—safely cached in each cell of the organism—that accrues the evolutionary benefits.

Setting a specific goal—as with artificial, guided evolution—means that there is a highly restricted *environment*; the basketball-player phenotype faces an environment in which it is demanded to perform a very specific task: playing basketball. Natural environments are not only much more complex but also highly *dynamic*; the phenotypes must face ever-changing circumstances.

Nature's open-endedness manifests itself not only at the phenotypic level but also at the genotypic level: Not only can the rules of the playground change, but so too can the rules for making players. The genome of a red ant is quite different from that of an orangutan (though as both are branches of the Tree of Life, they also bear many similarities). As we've seen, artificial-evolution scenarios to date are limited, being goal-oriented with but very little maneuverability in changing the genetic makeup. A bridge genome will always produce a bridge—perhaps a superb one at that—but never a skyscraper. Nature, though, can tinker with the genome, thus changing the underlying construction plan so as to produce entirely different beings, including skyscrapers (giraffes) and towns (ant colonies). This is a crucial aspect of her open-endedness.

We've seen how artificial evolution is used to design complex objects that stretch—or overstretch—our classical engineering techniques. The results are often quite impressive and at times those who use them are even reputed to cry out: "Wow, I'd have never come up with such a solution." But this is still at the level of evolving super bridges or superb basketball players; moreover, it might even be limited at that since an entirely novel bridge design or a new form of basketball player might necessitate genomic tinkering that is beyond the system's reach. Can something *truly astounding*—something entirely new—emerge out of an artificially set stage? In my mind this is one of our grandest challenges, and it may still be many years in the coming. I like to think of this challenge as that of building a system that Knocks Your Socks OFF. Following the time-honored tradition in computing science of coining acronyms, this might be dubbed the KYS OFF challenge, which leads me to wonder whether such a system would kiss *us* off.

As our artifacts become more and more complex, so does their design become more arduous. One way out is to employ the powerful process of open-ended evolution. But wait a minute. By definition, that would mean removing the designer from the equation! Then who controls the design process? Who's the boss? It seems that you can't have your cake and eat it too; something has to give. With guided evolution, the guide—or designer—maintains a great deal of control over the system, and, though the results obtained may often be overwhelming, the designer's socks will remain firmly in place. Open-ended evolution might indeed knock your socks off, but at the price of giving up some of the precious control that we've grown used to.

Strangely enough then, it is *less* design, meaning more open-endedness, that increases our design power. Uhm ... did I just say less design? Actually, you have to set up the stage so as to be more open-ended, that is, you have to design the system to exhibit less design! That's the essence of the KYS OFF challenge, which only Nature has met so far, but then again, she's been at it for the past 3.5 billion years. (While I've been concentrating my discussion of the open-ended versus the guided on evolution, this is by no means the only process of interest. Learning, for one, augments a system's open-endedness.)

Open-ended goes hand in hand with less control, though with the potential of more spectacular results. Parents usually want their children to grow up to be independent and able to think for themselves. But in many ways child rearing is open ended, with no guarantees: What if the child decides to be a rock star? (Result: horrified parents.) Or a doctor? (Result: delighted parents.) In Chapter 4 we discuss the application of biological processes, such as evolution and learning, in the field of adaptive robotics. We saw that one of the central goals is that of attaining more autonomous robots; I doubt, however, that we're ready to see them declare autonomy.

With an open-ended process not only do you not control the precise shape that the final outcome will take, but you're not even sure what this outcome will be. When we look at Nature's magnificent products with awe and with envy, we should always keep in mind the billions of years that their production necessitated. If you set off such a process and then patiently wait for a couple of billion years, you might find—*a posteriori*—lots of wonderful devices, such as eyes, toes, flowers, brains, and wings. You might be quite happy with this plethora of gadgets that will bring you fame and fortune. But this process is open-ended, which means you don't know in advance what the final products will be. In fact, it might not even get off the ground: It took Nature almost 3 billion years before things really started to pick up and the Tree of Life began to grow. What if it never gets off the ground? Or if you simply get tired of waiting?

The possibility of creating an artificial scenario in which open-ended evolution takes place is at the heart of Greg Egan's excellent science fiction novel *Permutation City*. Explaining to the researcher her mission, the protagonist says: "I want you to construct a seed for a biosphere.... I want you to design a pre-biotic environment—a planetary surface, if you'd like to think of it that way—and one simple organism which you believe would be capable, in time, of evolving into a multitude of species and filling all the potential ecological niches." Having succeeded in creating such a biospheric seed, evolution is then set loose in this artificial universe known as the Autoverse, to work its magic over the eons: "We've given the Autoverse a lot of resources; seven thousand years, for most of us, has been about three billion for Planet Lambert." And the outcome? "There are six hundred and ninety million species currently living on Planet Lambert. All obeying the laws of the Autoverse. All demonstrably descended from a single organism which lived three billion years ago—and whose characteristics I expect you know by heart. Do you honestly believe that anyone could have *designed* all that?" The answer is no; be it in an artificial or a natural world, open-ended evolution will knock your socks off.

If we give up our control, can't things get out of hand, leading to a system run amok? This is a tough question, which needs to be addressed on a case-by-case basis. We should come up with fail-safes (*à la* Asimov's three laws of robotics). We might well wish to place checks and bounds (for example, limit the robots' autonomy). We'd like to maintain the possibility of pulling the plug if things get downright ugly. But this issue is by no means an open-and-shut case: How does one juggle between control and autonomy, between guided and open ended? This issue will probably gain more prominence as our technology advances, enabling us to build systems that are somewhat less controlled, and less controllable.

Which brings me back to William Blake and the closing lines of *The Tyger*, where the "Could" of the first stanza has been conspicuously replaced by "Dare":

> *Tyger! Tyger! burning bright*
> *In the forests of the night,*
> *What immortal hand or eye*
> *Dare frame thy fearful symmetry?*

Dare we frame a tyger?

CHAPTER 13

The Scigineer

Science and engineering have traditionally proceeded along separate tracks. The scientist is a detective who's up against the mysteries of Nature: He analyzes natural processes, wishing to explain their workings, ultimately seeking to predict their future behavior. Scientists ask questions such as What goes on inside the Sun? And how long will it keep on burning? How does the weather system work? And how can we predict whether it will rain tomorrow or not? What are the fundamental physical laws that underpin the workings of the known universe?

The engineer, on the other hand, is a builder: Faced with social and economic needs, she tries to create useful artifacts. Engineers ask questions such as: How can we build a car with a cruising speed of 150 kilometers per hour, a fuel consumption of 20 kilometers per liter, and a price tag of no more than $8000? How do we design a computer chip that is twice as fast as the fastest extant chip? How can we build an autonomous lawn mower? "To put it briefly," wrote Lewis

Wolpert in *The Unnatural Nature of Science*, "science produces ideas whereas technology results in the production of usable objects." And if I may add my own little epigram, science is about making sense, whereas engineering is about making cents.

In a chapter entitled "Technology Is Not Science," Wolpert discussed the differences between the two, noting that technology is very much older than science and that science did almost nothing to aid technology until the nineteenth century. "Technology may well have used a series of ad hoc hypotheses and conjectures, but these were entirely directed to practical ends and not to understanding," he wrote. Humans have been able to construct artifacts—such as tools and arms—and improve their existence via agriculture and animal domestication thousands of years before the arrival of modern science (in the sixteenth and seventeenth centuries). Though engineers have only recently begun to put science to use, scientists had always relied on the existing technology. To quote Wolpert: "Science by contrast has always been heavily dependent on the available technology, both for ideas and for apparatus. Technology has had a profound influence on science, whereas the converse has seldom been the case until quite recently."

The emergence of technology long before science is not at all surprising. "The goals of the ordinary person in those times," wrote Wolpert, "were practical ends such as sowing and hunting, and that practical orientation does not serve pure knowledge. Our brains have been selected to help us survive in a complex environment; the generation of scientific ideas plays no role in this process." Thomas S. Kuhn considered science and technology in one of the most influential works in the philosophy of science, *The Structure of Scientific*

Revolutions, writing: "Just how special that community must be if science is to survive and grow may be indicated by the very tenuousness of humanity's hold on the scientific enterprise. Every civilization of which we have records has possessed a technology, an art, a religion, a political system, laws, and so on. In many cases those facets of civilization have been as developed as our own. But only the civilizations that descend from Hellenic Greece have possessed more than the most rudimentary science. The bulk of scientific knowledge is a product of Europe in the last four centuries. No other place and time has supported the very special communities from which scientific productivity comes." So perhaps we should count ourselves lucky to have science at all!

During the twentieth century the use of scientific knowledge in advancing the state of the art of our technology has picked up quite dramatically. Today all but the simplest artifacts rest on strong scientific foundations, everything from computer chips to automobile tires, through T-shirts, sugarless bubble gum, and space shuttles.

Science and engineering go hand in hand nowadays, both drinking from and helping to fill the other's fountain. We've seen how engineers not only apply our current scientific understanding of Nature in order to build better artifacts, but are indeed coming full circle, trying to make these objects more Naturelike. Biology serves as a source of inspiration, with processes such as evolution, learning, and ontogeny implemented in artificial media. Nature can even be directly co-opted for engineering purposes, as with the use of DNA molecules to solve problems in computing.

The betrothal of science and engineering, and the ensuing period of blissful courtship, have finally led, in my opinion, to marriage. I believe that recent years have seen the rise of a new kind of professional (and profession): the *scigineer*, a combination of both scientist and engineer, holding a test tube in one hand and a proverbial slide rule in the other.

What is a scigineer? Let me go about explaining this by way of example. In Chapter 2, we saw how computer programs in the form of trees can be evolved, noting that evolution tends to produce "spaghetti" programs: huge trees with lots of weird branches and offshoots. If the program works to your satisfaction, you can of course simply go ahead and use it; if you want to understand what makes it tick, though, then you're in a position that's rather like that of a biologist trying to decode our own program (the human genome). We even noted that when you delve into these evolved programs, you frequently find loads of "junk": computer code that is of no use at all, a situation which is similar to Nature. Our genomes also contain junk code: unused portions of our DNA program.

The scigineer has two hats—that of a scientist and that of an engineer—which she constantly alternates. First, she puts on the engineer's hat, picks up her slide rule, and sets the stage, say, for the evolution of computer programs; then, she puts on the scientist's hat and the white coat, setting out to analyze the creatures (programs) that have emerged in her artificial universe.

The robots of Chapter 4 also constitute a case in point. They are artifacts created by the scigineer, who subjects them to an environment in which they evolve and learn. We saw how they can come to avoid obstacles, but exactly how do they accomplish this? Though we're talking about an artifact—an object created by humans—it has evolved into something that we do not fully comprehend. Even though as stage designers we seem to have a privileged position, the actors have taken their own routes so as to better themselves. The scigineer must now take out his scientific toolbox in order to analyze this little robotic creature, just as a scientist analyzes a cockroach. Though such a current-day robot is still a far cry from a cockroach, it's already complex enough to require the donning of a white coat.

Let me give you another well-known example, that of the Tierra world. Tierra is a virtual universe—embedded within a computer—that was set up in an attempt to explore the idea of open-ended evolution. It comprises computer programs that can evolve; unlike those of Chapter 2, however, where an explicit goal (and hence fitness criterion) is imposed by the user (for example, compute taxes), the Tierran creatures receive no such guidance. Rather, they compete for the natural resources of their computerized environment: time and space. You may remember from Chapter 6 that a standard computer consists of two major elements, the processor that actually runs the program and the memory, the storehouse that acts as a repository for programs. These two components represent Tierra's natural resources, and—just as in Nature—they are limited: The processor can run only one program at a given moment, and the memory can contain no more than a certain number of programs. This gives rise to a fierce battle for survival, the Tierran creatures having to vie for the processor's precious time and for a place in the jungle known as memory. Failure means death: A program that is unsuccessful in procuring these resources disappears from the evolutionary stage.

Tierra was invented not by a computing scientist but by an ecologist, Thomas Ray, who had worked for years in the Costa Rican rain forest before turning from natural evolution to digital evolution. Ray inoculated his Tierran world with a single organism—a self-replicating program called the "ancestor," which was able to co-opt the processor to produce copies of itself elsewhere in memory. This organism, a program written by Ray himself, was the only engineered (human-made) creature in Tierra. The replication process is not perfect: Errors, or mutations, may occur, thus driving the evolutionary process. Ray then set his system loose and witnessed the emergence of an ecosystem in a bottle, right there inside his computer, including organisms of various sizes, and such beasties as parasites and hyperparasites. Ray wrote that "much of the evolution in the system consists of the creatures discovering ways to exploit one another. The creatures invent their own fitness functions through adaptation to their biotic environment."

Large programs such as the ancestor have several instructions that form part of their "body"; these program instructions are used to copy the organism from one memory location to another, thus effecting replication. The evolved parasites are small creatures (programs) that use the replication instructions of such larger organisms to self-replicate. In this manner they proliferate rapidly in the memory jungle without the need for the excess replication code. As in Nature, the evolved ecology exhibits a delicate balance: If all large creatures were to disappear, then the parasites would die, having no replication code to appropriate. Tierra had even managed to outdo its creator, who wrote: "Comparison to the creatures that have evolved shows that the one I designed is not a particularly clever one."

Ray first engineered this world, which he then proceeded to analyze as a scientist: "Trained biologists will tend to view synthetic life in the same terms that they have come to know organic life. Having been trained as an ecologist and evolutionist, I have seen in my synthetic communities, many of the ecological and evolutionary properties that are well known from natural communities." (If you're interested in learning how the humble Tierran beginnings ultimately lead to the rise of the "TechnoCore" artificial intelligences [AIs], I recommend *The Rise of Endymion*—the final volume of Dan Simmons's *Hyperion* tetralogy.)

In April 1998, while leafing through the weekly issue of *Science*, I was surprised to find two out-of-the-ordinary articles. *Science*, one of the top two scientific journals (the other being *Nature*), publishes almost exclusively hard-core scientific papers in physics, chemistry, biology, and the like. If your paper is good enough to grace *Science*'s pages, then it's probably about the natural world—the object of scientific study. Yet in browsing this particular issue, I suddenly came across a couple of articles that dealt with an artificial world created entirely by humans: the World Wide Web. One article looked into the efficiency of search tools, while the other studied patterns of behavior as information foragers move from one hyperlinked document to the next. It's almost as if we were talking about a tropical jungle.

This is a cogent example of scigineering that is totally unrelated to biology or biological inspiration. Here is a universe created entirely by humans, which has become so complex—much more so than a car or an elevator—and so *interesting in and of itself* that it merits the attention of scientists—and the consecration of *Science*. We've engineered the World Wide Web, and then we turn to study this brave new world. The era of scigineering is upon us.

The rival journal, *Nature*, waited until August 1999 to finally "give in." In an article entitled "Genome Complexity, Robustness and Genetic Interactions in Digital Organisms,"

Richard Lenski, Charles Ofria, Travis Collier, and Christoph Adami explored the effects of genetic mutations in both simple and complex digital organisms, which inhabited the artificial, Tierra-like world called "Avida." Commenting on their work, Inman Harvey from the Centre for the Study of Evolution at the University of Sussex cautioned that "considerable debate can be expected before a consensus is reached on just what is necessary for results from a synthesized world to be seen as relevant to the natural world." The scigineer might study her world and glean much about it, but she must be cautious in applying her conclusions to the world at large.

The scigineer has one up on the scientist in that he can render his world easier to analyze, whereas a scientist must make do with what Nature affords him. Evolutionists would love to have the *entire* Tree of Life at their disposal, including all the lost species, yet this is but wishful thinking; geological reality, alas, is harsher on them, revealing but bits and pieces of the whole story. With artificial worlds, though, wishes are granted: You can easily save the entire evolutionary history of your artificial creatures to later analyze it at your leisure.

Remember from Chapter 12 how the protagonist of *Permutation City* describes the result of the Autoverse experiment—the result of billions of years of evolution in this artificial world? He says: "All *demonstrably* [my emphasis] descended from a single organism which lived three billion years ago...." On this artificial planet, one can *demonstrate* that all the organisms have descended from a single origin since the entire evolutionary trace is available. The scigineer might not possess perfect knowledge of his engineered world, but he at least has the power—unlike scientists—to render his analysis job easier.

In reminiscing about his illustrious career, Isaac Newton remarked: "I do not know what I may appear to the world; but to myself I seem to have been only like a boy playing on the seashore, and diverting myself in now and then finding a smoother pebble or a prettier shell than ordinary, whilst the great ocean of truth lay all undiscovered before me."

We're no longer content to walk the shores of Nature's oceans of truth, finding whatever pebbles may have been laid down for us. We're now creating new oceans, and with them we beget new shores to walk.

CHAPTER 14

A Life of Its Own

"**H**e wanted to dream a man: he wanted to dream him with minute integrity and insert him into reality." This was the goal of the silent man who came from the South, in Jorge Luis Borges's short story "The Circular Ruins." From Pygmalion, Frankenstein, and the Golem to *Star Trek*'s Lieutenant Commander Data, the dream of administering the breath of life has fascinated humankind since antiquity.

We've seen how human-made systems can be made to evolve, to learn, to adapt, and to develop, as well as to exhibit a host of other characteristics that are usually not associated with machines, but rather with living beings. Can our creations one day take on a life of their own? This question moved from the realm of science fiction to that of science with the advent of the field known as *artificial life*. The term was coined by Christopher G. Langton, organizer of the first artificial-life conference, which took place in Los Alamos in 1987.

"Artificial Life," wrote Langton (in the proceedings of the second conference), "is a field of study devoted to understanding life by attempting to abstract the fundamental dynamical principles underlying biological phenomena, and recreating these dynamics in other physical media—such as computers—making them accessible to new kinds of experimental manipulation and testing." While biological research is essentially *analytic*, trying to break down complex phenomena into their basic components, artificial life is *synthetic*, attempting to construct phenomena from their elemental units, and, as such, adding powerful new tools to the scientific toolkit. This is, however, only part of the field's mission. As put forward by Langton "In addition to providing new ways to study the biological phenomena associated with life here on Earth, *life-as-we-know-it*, Artificial Life allows us to extend our studies to the larger domain of the 'bio-logic' of possible life, *life-as-it-could-be*, whatever it might be made of and wherever it might be found in the universe."

Before talking about *artificial* life, shouldn't we try to define what life is? Well … no. I'll steer clear of this issue since it is in fact quite a controversy in science; as things stand today, there is no agreed-upon scientific definition of *life*. For now, we'll just have to accept its being one of those "you-know-it-when-you-see-it" qualities: Your dog is obviously alive, while your washing machine is obviously not. The question is, then, can we create something that is "obviously alive"?

This question seems reasonably clear, or is it now? You'd think that it's the "life" part of "artificial life" that eludes our definition. Well, there's a further subtlety: What exactly does the "artificial" part mean? If you look up "artificial" in the dictionary (Merriam-Webster online), you'll find a number of definitions. So let's see which one sits well with "artificial life." Artificial might mean "lacking in natural or spontaneous quality < an *artificial* smile > < an *artificial* excitement >."

This can't be it. An extraterrestrial might be unnatural and unspontaneous, and yet is obviously alive; artificial life cannot be about life that lacks in natural or spontaneous quality. What about "imitation, sham < *artificial* flavor >"? This is no good: By definition our putative "artificially alive" creature is going to be an imitation in some sense; the point is, in what sense, and how good an imitation? ("That's not a real dog? I never would've guessed in a million years!") Saying that artificial life is synonymous with imitation life doesn't get us very far. We're obviously trying to imitate life, in the proverbial "imitation is the sincerest flattery" sense.

I'm not merely engaging here in armchair philosophy, but rather trying to arrive at the essence of what "artificial life" means. What seems to be missing in the two definitions of the previous paragraph is the *creation* aspect. So let's try what is actually the first definition appearing under "artificial": "man-made < an *artificial* limb > < *artificial* diamonds >." Ah! now we're cooking. This seems to be the right one. It accords perfectly with the definition given by Langton in the proceedings of the first artificial-life conference: "The study of man-made systems that exhibit behaviors characteristic of natural living systems."

Artificial life is thus life created by humans rather than by Nature. Simple. Well... I hate to be so fussy, but "human-made" can mean at least three different things. One way to create life is through the union of a male, known as "daddy," and a female, known as "mommy," thus giving life to a male or a female known as "baby." You might be frowning now, thinking to yourself that it's rather silly of me to even mention this since this is quite obviously *not* artificial life. This is the natural way of creating life, whatever that may mean. But then again, what about artificial insemination? This involves the introduction of semen into the uterus or oviduct by other than natural means, yet no one would claim that this produces artificial babies. Nonetheless, there is a definite intervention by humans, thus rendering this process somewhat less than 100 percent natural.

We often invoke the term "human-made" when speaking of objects such as cars. This image of artificial life might involve some kind of production line, where heads, arms, feet, and torsos are assembled into complete beings, after which the proverbial switch is pulled, thus breathing life into them. (The assembly line need by no means produce but humanoid life; it could in fact produce a range of beings, from artificial bacteria to artificial whales.) This is the most common image where fiction is concerned (Victor Frankenstein creating a humanoid monster, for one).

There is yet a third way by which life may be created by humans: through the process of evolution, and most likely open-ended at that (as we discuss in Chapter 12). This raises an interesting question: While we may sow the seeds of life, setting off such an open-ended process, whatever emerges—numerous generations later—might be far removed from our original design; just how "human-made," then, is this form of life?

My intention in the somewhat philosophical discussion above has been to show you just how intricate this seemingly simple term—artificial life—really is. The concept of "artificial" is quite elusive where life is concerned, and even if we agree on emphasizing the "creation" aspect, there are a number of fundamentally different modes of creation.

Artificial life might in fact be an oxymoron. After all, how can life be artificial? If something is truly alive—assuming we can somehow agree on this fact—then what's artificial about it? Even if we take what could arguably be considered the most artificial route of creation, that of the assembly line, once we're done, the creature is no longer artificially alive; it's alive—period. This takes us right back to Langton's definition of artificial life, life-as-it-could-be, "whatever it might be made of and wherever it might be found in the universe." Whether flesh-and-blood, man-woman-made, or nuts-and-bolts, factory-made, life is life. Perhaps rather than speak of artificial life, which is somewhat problematic, we should talk about life created by humans. In fact, even "created" might be too strong a word (think of the evolutionary scenario, for one). Let's settle for human-induced life. This emphasizes the relevant difference between Nature and humans, namely, the *manner* by which life arrives on the scene; the end result though is—in both cases—*bona fide* life.

Life may be many things: perhaps "a tale told by an idiot, full of sound and fury, signifying nothing" (Shakespeare), or maybe "colour and warmth and light, and a striving evermore for these" (Julian Grenfell), or indeed "a glorious cycle of song, a medley of extemporanea" (Dorothy Parker). At the heart of artificial-life research lies the belief that whatever life is, it is not about carbon; life is not about the medium but about the mediated. It is a *process* that we do not yet understand in full, but which we may nonetheless be able to create, or perhaps we should say re-create: After all, Nature has beaten us to it.

Let's get down to Earth now and consider some of the issues involved in inducing life. As I discussed above, there are at least three ways of going about this. We might imagine some far-future extension of current medical practice (such as artificial insemination) that will result in a new form of life. Since this involves many technical biological and medical details, I think I'll leave it at that for the present discussion.

The second way to induce life is to produce a full-blown living being. As I briefly mention in Chapter 11, there is much ongoing research on mimicking Nature's gadgets, building such devices as eyes, ears, and hearts. While many of these are intended to serve as prostheses for humans, some are also used in robots. Perhaps at some point in the future we'll be in possession of enough parts to construct an entire being. This might in fact come sooner rather than later: While speaking of "inducing life" usually tends to evoke in us images of humanoid life, let's not be *Homo sapien* chauvinists. As I've mentioned time and again, constructing the equivalent of even a single-celled organism would be a huge achievement (not to mention a beetle or a fly), and this might come about sooner than we expect. (John Wyndham's short story *Female of the Species* provides an amusingly gruesome vision of this production-line scenario. When visited by two inspectors of the Society for the Suppression of the Maltreatment of Animals, Doctor Dixon—the Frankenstein-like protagonist—explains: "The crux of this is that I have not, as you are suspecting, either grafted, or readjusted, nor in any way distorted living forms. I have *built* them.")

And then there's the third way of inducing life, by creating the necessary conditions for open-ended evolution to take place. In Chapter 1 we note that evolution rests on four principles:

- Individual organisms vary in their viability in the environments that they occupy.
- Individual variation is heritable.
- Individuals tend to produce more offspring than can survive on the limited resources available in the environment.
- Those individuals best adapted to the environment are the ones that will survive to reproduce in the ensuing struggle for survival.

The second and third principles—heritability and the production of offspring—conceal an even more fundamental property: self-replication. We mention this in relation to the Tierra world (Chapter13), which was inoculated with an ancestral (computer-program) creature able to self-replicate, creating copies of itself in the computer's memory. Self-replication was a necessary precursor to the appearance of life on Earth: The first self-replicating molecules had appeared about 3.5 billion years ago, a humble beginning that marked the onset of the evolutionary avalanche that gave rise to life.

Tierra inventor Thomas Ray wrote, "I would consider a system to be living if it is self-replicating, and capable of open-ended evolution." (The Tierran world was in fact set up to discover not *how* self-replication arrives on the scene, but what happens *after* it does, namely, how a diverse ecosystem comes to evolve.)

The study of self-replicating structures in human-made (or human-induced) systems began in the late 1940s, when John von Neumann—one of the twentieth century's most eminent mathematicians and physicists—posed the question of whether a machine can self-replicate (produce copies of itself). He wrote that "living organisms are very complicated aggregations of elementary parts, and by any reasonable theory of probability or thermodynamics highly improbable. That they should occur in the world at all is a miracle of the first magnitude; the only thing which removes, or mitigates, this miracle is that they reproduce themselves. Therefore, if by any peculiar accident there should ever be one of them, from there on the rules of probability do not apply, and there will be many of them, at least if the milieu is reasonable."

Von Neumann was not interested in building an actual machine, but rather in studying the theoretical feasibility of self-replication from a mathematical point of view. He succeeded in proving (mathematically) that machines can self-replicate, laying down along the way a number of fundamental principles involved in this process. During the decade following his work (the 1950s), when the basic genetic mechanisms had begun to unfold, it turned out that Nature had "adopted" von Neumann's principles. (It is quite fascinating to see how his results predated the actual biological findings.)

The study of self-replication has been taking place for more than half a century. Much of this work is in fact quite separate from artificial life and is motivated by the desire to understand the fundamental principles involved in this process. This research might better our understanding of self-replication in Nature, as well as find many technological applications. There is much talk today of nanotechnology, where self-replication is of vital import. You'd like to be able to build one tiny machine, which would then sally forth and multiply. For example, you'd inject a small nanomachine into your body to fight off some mean virus, and this nano-machine would be able to self-replicate, thereby increasing the size of your internal army. One of my favorite application examples is the self-replicating lunar factory, which is not drawn from some science fiction novel but was actually proposed by NASA researchers in 1980. Imagine planting a "seed" factory on the moon that would then self-replicate to populate a large surface, using local lunar material. This multitude of factories could manufacture necessary products for lunar settlers or for shipping back to Earth. And all you have to do is plant the first one.

On our way to inducing life, self-replication is of crucial import. We know a bit more about this issue today than we did 50 years ago, though there is still no lack of unanswered questions, which is music to researchers' ears. The next item on our life-inducing agenda is trying to come up with an open-ended context for our self-replicating critters (just as Ray set out to do with his Tierran world); this issue has both genotypic and phenotypic aspects (Chapter 12).

The phenotypic aspect of open-endedness concerns the environment. The grand challenges posed by an open-ended environment vis-à-vis its inhabitants are to be able to move around and to sense the surroundings to a degree sufficient to achieve the necessary maintenance of life and reproduction. What seems to us to be really difficult—like playing chess— may in fact be quite easy once this essence of being and re-acting is available. Remember, elephants don't play chess (Chapter 11).

Chess is in fact quite an instructive example. It has been one of the holy grails in the field of artificial intelligence since the 1950s. In those early days a number of researchers had managed to come up with programs that were able to play a decent game, winning against average human players (though they were easily beaten by chess experts). The ruling opinion at the time was that very soon there would be a chess ma-chine able to beat any human player. The problem, though, turned out to be harder than believed, demonstrating what is known as the fallacy of the first step: It's easier to go from ignorance to mediocrity than it is to go from mediocrity to excellence (think of the difference between playing tennis and playing tennis *well*). Mediocre chess-playing computers were available as far back as the 1960s, though only very recently has a computer (IBM's "Deep Blue") been able to beat a world champion (Garry Kasparov).

It took 40 years to come up with a good chess-playing computer, and, frankly, chess is easy! I'm not saying it does-n't require a form of genius to excel at the game, nor am I belittling the arduous task faced by the designers of a chess-playing machine; I'm referring to the facileness of the chess *environment*; it is yet another illustrative example of non-open-endedness. Chess is defined by a very small number of well-known, fixed rules, and there's really no dynamically changing environment to speak of (in this sense it is similar to our basketball environment of Chapter 12).

How do we increase the open-endedness of the environment? Thomas Ray took what is perhaps the first shot at providing an answer to this question with his simulated Tierra world. Another possibility is to subject our critter to the most complex environment known to date: ours. This is the route taken by adaptive-robotics researchers. In Chapter 4 we saw how real robots are subjected to a real-world environment (as opposed to simulated robots in a simulated computer environment); for this reason the approach is also known as *situated* or *embodied* robotics.

An argument that is often raised against embodied robotics is that it is too costly and too slow. You'd be much better off running the evolutionary process in a simulated environment within the computer, plucking out only the end result— the best simulated robot that has evolved—and implementing it in the real world. The problem is that often this does not work: When you go from the simulated to the real, the robot no longer functions properly. Our environment is full of many hidden complexities that often escape our notice, rendering it very hard to implement them in a computer; it's just plain easier to use the real world.

Nature's open-endedness manifests itself not only at the phenotypic level but also at the genotypic level; she can tinker with the genome so as to produce entirely novel designs, which give rise to new phenotypes better able to rough it. This point has also been receiving increased attention of late: How can we set up an evolutionary scenario in which fundamental genomic changes can occur? For example, in Chapter 4 we evolved only the *behavior* of the robots, their small, neural-network brains. Their bodies, on the other hand, did not change at all, which is nothing like natural evolution. Nature possesses the ability to bring about not only behavioral changes but also morphological modifications in her creatures. A number of researchers have recently begun looking into the possibility of doing this for robots as well, evolving both behavior and morphology. While quite rudimentary at the moment, this is yet another step toward increased open-endedness.

Next to the natural world a new universe has sprung up in the past few years, which is both complex and open-ended: the Internet. It is evolving at a breathtaking speed, already exhibiting enough complexity to merit the attention of scigineers. This is not surprising. The Internet's evolution is mediated by self-proclaimed intelligent beings known as *Homo sapiens*. This process is more akin to Lamarckian evolution, where a beneficial survival trick can be immediately incorporated within the evolving population.

Will we someday see the rise of *network life*? Even as you read these lines, there are thousands and thousands of small programs—known as agents—roaming the network, seeking to find information that will appease their human masters. Currently they are quite limited, lacking in both intelligence and autonomy. Little by little, though, they might develop into more autonomous critters. This might come about by employing some of the techniques we discuss in this book, giving rise to what I call *Egents*, for Evolving Agents, and *double AAgents*, for Adaptive Agents. These agents will be denizens of the network universe, whereas we will not; it is they who will be in their element. We may have built the house, but we are not the ones living in it. I just hope those double AAgents will work for *you*, their master, and not for some unknown party behind the cyber curtain.

The borders between the living and the nonliving, between the Nature-made and the human-made appear to be constantly blurring.

As in dreams.

The silent man who came from the South eventually succeeded in dreaming a man and inserting him into reality. And what became of the dreamer? "With relief, with humiliation, with terror, he understood that he too was a mere appearance, dreamt by another."

Sweet dreams.

CHAPTER 15

From Adaptive
to Zestful

Accurate	Justifiable	Serial
Binary	Knotty	Testable
Consistent	Logical	Universal
Deterministic	Mechanistic	Verifiable
Electronic	Numerical	Will-less
Formal	Optimal	Xeroxed
General-purpose	Precise	Yielding
Humdrum	Qualifiable	Zippy
Ignorant	Rigid	

The preceding list includes but a sampling of descriptors that have been used in conjunction with computing since the rise of the modern-day computer in the 1940s. The past few years have seen a plethora of novel terms that have entered the computing dictionary. While many of these ideas find their roots in the early days of the field, with pioneering work carried out by researchers in the 1940s, the 1950s, and the 1960s, their impact is quite recent. Indeed, there seems to be a bombardment of new terms, novel buzzwords humming all around us.

For my grand finale I'd like to present you with twenty-six of the newer entries in the ever-expanding computing dictionary. Some of these serve to summarize central concepts that we've discussed in this book, while others appear here for the first time. I'm not setting out to provide you with a complete dictionary, else this chapter would turn into a book in its own right. I just wish to shine a light—a flashlight—into some interesting regions of the current and future computing landscapes.

Samuel Johnson wrote: "Dictionaries are like watches; the worst is better than none, and the best cannot be expected to go quite true." I hope that if mine is anything like a watch, then it's like a self-repairing watch at that.

ADAPTIVE We come full circle, once again meeting our cherished friend of the first chapter, which we've kept running into throughout this book. By now we know that adaptation refers to a system's ability to undergo modifications according to changing circumstances, thus ensuring its continued functionality within the environment in which it operates. Adaptation can take on many guises and can arise as a consequence of several processes. Computing systems can be rendered more adaptive by embodying processes inspired by Nature, by judicial recourse to network resources, and through improved human-computer interaction. Adaptiveness is thus like a silken thread, intimately woven into the pages of our dictionary.

BIOLOGICAL Again we encounter what is by now an old friend. As we've seen throughout, the influence of the biological sciences in computing has been gaining prominence, slowly but surely inching its way toward the mainstream. The motivation stems from the fundamental observation that living organisms exhibit several desirable characteristics that have proven difficult to realize using traditional engineering and computing methodologies.

COMMUNICATIVE Just to keep you on your toes, I've opted here for "communicative" rather than the "complex" of the first chapter. Gigabytes of text have been written on the emergence of computer networks and their pivotal, increasing influence on our daily lives. It is interesting to note the shift from one C to another—from computation to communication. When digital computers had first appeared over 50 years ago, they were viewed exclusively as computing machines. With the spectacular rise of the Internet, we are witness to a shift. The computer is no longer regarded solely as a means for calculation, but primarily as a machine that provides new ways for people to communicate with one another. *I link, therefore I am.* And if I am, why not compute? Indeed, more recently we've come full circle, going from communication back to computation. In a recent article entitled "The Challenge of Tamperproof Internet Computing," Edmund Ronald, from the Ecole Polytechnique in Paris, and I wrote: "The prevailing tendency to exploit computation to achieve better communication is swiftly redefining the global information ecology by making every published item instantly visible in any part of the world. However, the inverse trend—harnessing global communication to achieve more powerful computation—is also developing before our eyes.... Researchers have proposed various schemes to transform the Internet into the 'Interputer.'"

DYNAMIC By this term I refer to systems that are marked by continuous and productive activity or change, which are thus able to handle complex environments. In other words, the term "dynamic" refers to *dynamic* systems that are able to cope with *dynamically* changing circumstances. Much of this book has been about boosting the dynamism of computing systems.

EVOLVING Whether guided or open-ended, purposeful or purposeless, Darwinian or Lamarckian, the importance of the process known as *evolution* is on the rise. As enunciated by Charles Darwin in the closing lines of the *Origin of Species*: "There is grandeur in this view of life, with its several powers, having been originally breathed into a few forms or into one; and that, whilst this planet has gone cycling on according to the fixed law of gravity, from so simple a beginning endless forms most beautiful and most wonderful have been, and are being, evolved."

FUZZY Computers are beginning to understand the virtues of being fuzzy. Fuzzy logic for one (Chapter 5) allows computing machines to go from "if the room temperature rises above 27° Celsius, then increase motor output by 34 percent" to "if the room is hot, then substantially increase motor speed." Fuzziness, though, is more than that: It also refers to a "messier" way of computing, more akin to how things work in Nature. If life can be messy, why can't computers? (Note that while life is messy, the converse is not true: Lucky for us, not every mess is alive.)

GROWING One prime difference between machines and living organisms is that the former are constructed full blown, while the latter grow. Suppose we were able to grow machines? We saw in Chapter 7 how you can plant the embryonic seed of a multicellular machine, which then develops and grows into a full-fledged organism. As you may remember, the electronic substrate in this case is a configurable processor (Chapter 6), that is, we assume the existence of a large chessboard of cells; no material is actually created. A number of researchers recently have been looking into the possibility of having true material growth: growing nerve cells and cables on a silicon substrate, thus obtaining integrated neuron-silicon systems that act as artificial neural networks (Chapter 3), or growing wires in a computer chip rather than connecting them manually. While this may seem hard to swallow at first, it definitely grows on you.

HEALING One possible path to attaining more resilient computers is by furnishing them with healing, or self-repair, capabilities. We've seen two ways in which this can be accomplished. One is based on the immune system of living beings, which is capable of learning, recognizing, and above all eliminating foreign bodies that continuously invade the organism (Chapter 8). Another artificial system that heals is the embryonic circuit of Chapter 7, which is based on the multicellular structure of complex organisms and on the way they come to be—the developmental process known as *ontogeny*.

ICONIC Over the past decade computing environments have grown to be more and more iconic, or pictorial. Most major computing systems in use today rely to a great extent on the manipulation of visual, iconic cues. Since human beings are predominantly visual animals, such iconic representations facilitate human-computer interaction. To date, there is still much to be wanted in this area, and major innovations are still in the waiting. One oft-mentioned idea is that of virtual reality, where the user is immersed in a three-dimensional world. The diffusion of this technology to the point of ubiquity may dramatically change the way we view computers.

JAVANESE The recent emergence of the Java programming language has brought along a major conceptual innovation to the Internet. A program written in a standard programming language (such as Pascal, BASIC, or FORTRAN) often requires tedious work by the programmer if he wishes it to run on different types of computers (for example, an IBM PC, an Apple Macintosh, and a Unix workstation). Java has been specifically designed for use as an Internet language, with Java programs able to run on any kind of computer. No longer are we confined to the perusal of passive information over the Net; we now have access to active information as well in the form of programs.

KNOWLEDGEABLE In an article on the future of higher education, Peter J. Denning distinguished between data, information, and knowledge, writing: "Data are symbols inscribed by human hands or by instruments.... Information is the judgment... that given data resolve questions, disclose or reveal distinctions, or enable new action. In other words, information is data that makes a difference to someone... Knowledge is the capacity for effective action in a domain of human actions." Today, we are witness to an unprecedented information (and data) explosion, though we still lack the tools that will allow us to gain knowledge.

LEARNING The ability to learn throughout one's lifetime via interactions with the environment is exhibited by most higher forms of living organisms and is most pronounced in human beings. This is yet another mode of adaptation, which has served as an inspiration for the development of artificial neural networks (Chapter 3). Such networks are able to solve difficult problems from examples, a radically different approach from that of directly programming the desired behavior.

MOLECULAR Computing devices keep getting smaller and smaller; there are millions of tiny widgets—known as transistors—in modern-day chips. Currently we just keep on stretching the existing technology, known as VLSIC (very large scale integrated circuits). At some point though (probably within 10 to 15 years) this technology will have to be supplanted by an entirely new one. As we saw in Chapter 9, molecules—and molecular biology—might provide us with the needed edge; molecular computers may be part of our future.

NATURAL In *Man and Superman,* George Bernard Shaw wrote: "The reasonable man adapts himself to the world; the unreasonable one persists in trying to adapt the world to himself. Therefore, all progress depends on the unreasonable man." These words capture our changing view of computers: Rather than adapt humans to machines, we should aim for the opposite, having machines adapt to us. We want to humanize computers instead of mechanizing humans. In practice, this involves a more human-centric approach, where the goal is no longer merely the construction of the best possible computer (the fastest, the largest, the most energy-efficient). Rather, there is an additional emphasis on the computer's use by humans, which ultimately implies a more *natural* way of interfacing. Natural to us, that is.

ORIENTED In discussing the future of computing, Richard W. Hamming recently wrote: "While one reads about transistors that will change state with the passage of a single electron (maybe), further reductions in line widths, occasional articles on thermodynamically reversible computers (to absorb the heat), and for chemical and molecular computers, one has to wonder about the delivery of these innovations for general-purpose computing within the next fifty years." Remember from Chapter 6 that a general-purpose computer is able to run any correctly written program, though it is neither the most efficient nor the fastest at that. While the goal of general-purpose computing is laudable, there is also a vast market for *oriented* computers. Though these may be able to operate within but a narrow domain or perform only a limited number of functions, they may do so quite efficiently at that and thus be advantageous. Such oriented computers are already upon us today (in most of your home appliances for one), and their use will probably increase dramatically with the advent of ubiquitous computing (discussed below).

PARALLEL Parallel computing refers to the use of a large number of computers that work together to solve a given problem. Though there has been much research in this domain since the 1960s, and there seemed to be much promise, it turned out that such computers are very hard to program. Today, parallel machines are still confined to small niches, with most programmers continuing to use standard (nonparallel) computers. Considering this lack of success, parallel-computing pioneer Michael J. Flynn wrote that "we significantly underestimated the difficulty in achieving the performance speedup expected from parallel processors." More recently, we are witness to a resurgence of novel, massively parallel techniques, such as cellular computing, molecular computing, and quantum computing. Flynn may thus be right in stating that "Parallel processors were the future . . . and may yet be."

QUANTUM Recently there has been great interest in machines that can take advantage of quantum physics. The reason that this domain stirs so much excitement is the possibility it raises of ultimately building a so-called quantum computer, which would exhibit *quantum parallelism*. Roughly speaking, a quantum system is not in a single classical state, but in a quantum state, which consists of a superposition of many classical states. With a classical computer, the basic element of computation is a bit—a binary digit—which may be *either* 0 or 1 at a given moment. With a quantum computer, the basic element is a quantum bit—or qubit—which can be in a superposition of these two states; in a sense it is *both* 0 and 1—this is the crux of quantum parallelism. Quantum computers are theoretically not only parallel but indeed *highly* parallel, which means that they could tackle hard problems (such as the traveling salesperson problem of Chapter 9). At least this is the dream; in reality, there are many obstacles yet to overcome, both theoretical and practical, and the jury is still out on whether the dream could someday be realized.

ROBOTIC As I note in Chapter 4, one of the starkest unfulfilled promises to date of computing science and engineering has to do with robotics. There is a second wind, though, currently sweeping the robotics shores, and the years to come might finally see the dream realized. Someday, we may find ourselves living side by side with a new species, one created by humans.

SCHOLASTIC The use of computers in educational institutions, ranging from kindergarten to college, is a trend that is likely not only to continue but indeed to accelerate. The crux of this issue lies in the shift from using computers in secondary roles (for example, secretarial or library related) to the primary role of teaching; this may well change the teaching profession. The historical role of the teacher has been that of a conveyor of information, transmitted to a (usually) large class at a uniform pace, using simple teaching aids. With the entrance of computers into the classroom, this doctrine no longer holds, and the teaching task becomes a joint human-computer venture. Students can learn at their own individual pace, using state-of-the-art learning aids. The (human) teacher's task is now different: He is to *guide* and *counsel* each student on her own individual path, making sure that progress is constantly being made. Will human teachers disappear altogether? Not to worry. It is unlikely that computers will supplant them since humans provide a crucial element: the human touch.

TOLERANT Typically, computers have exhibited very little tolerance to demands that fall outside their narrow functional specification. At the machine level, such demands can result from component failure, be it software or hardware. We would therefore like computers to be more fault-tolerant in general. Another form of tolerance relates to the interaction of humans and computers. We would like to imbue computers with the ability to tolerate ambiguity, fuzziness, and lack of completeness, all of which are characteristic of human interactions. We are all too aware of the frustration resulting from the need to be precise, exhaustive, and flawless when interacting with computers, qualities that are in stark contrast to our normal mode of thinking and interacting.

UBIQUITOUS Mark Weiser and John Seely Brown, in an article discussing what they call "calm technology," divided the age of computing into three eras: (1) the era of the mainframe, when computers were few and expensive, (2) the era of the personal computer, when computers were sufficiently numerous so that each person could have one, and (3) the coming era of *ubiquitous computing*, when computers will vastly outnumber people. They wrote that this era "will have lots of computers sharing each of us." They argued that the social impact of this trend may be analogous to two other technologies that have become ubiquitous: writing and electricity. These are so commonplace and unremarkable that we forget their huge impact on everyday life. Computing may join these two technologies in becoming ubiquitous.

VIRTUAL The impact of this term was nicely conveyed by Abbe Mowshowitz, in an article on virtual organization: "The popularity of *virtual reality* has led to the coupling of 'virtual' with any number of nouns to characterize the changes brought about by networking in various familiar settings." Indeed, one finds virtual offices, classrooms, organizations, and so on. At times it seems that only we humans are the real, nonvirtual element—at least for now.

WEARABLE One aspect of ubiquitous computing is that of wearable computers. Steve Mann, a pioneer in this domain, wrote: "Let's imagine a new approach to computing in which the apparatus is always ready for use because it is worn like clothing... Clothing is with us nearly all the time and thus seems like the natural way to carry our computing devices... our computer system will share our first-person perspective and will begin to take on the role of an independent processor, much like a second brain—or a portable assistant...."

XENOPHOBIC Issues of security and privacy have been receiving wide attention in the past few years, in the wake of the Internet's ascent. Metaphorically speaking, one can view this as the xenophobic side of computing, where some degree of stranger wariness is practiced.

YAHOOD Yahoo is the well-known Web index (www.yahoo.com) aimed at facilitating the life of the modern-day Web surfer. Yahood thus represents the dire need to have effective means of searching within the oceans of information—and finding that which we seek. As written by T. S. Eliot in *The Rock*: "Where is the wisdom we have lost in knowledge? Where is the knowledge we have lost in information?"

ZESTFUL I end with this term to emphasize the fact that much work is being carried out today to ameliorate human-computer interaction. This will result not only in improved efficiency, but perhaps, more importantly, in an agreeable and enjoyable rapport. Let's not forget that as humans we have *feelings.* In an article entitled "Why Must Computers Make Us Feel Blue, See Red, Turn White, and Black Out?" Edmund Ronald and I wrote: "Humans and computers are two alien races that often fail to communicate smoothly (or at all), despite one race having created the other. In their wisdom, computing professionals have not failed to note this alienation problem... To make clear that this is a serious business, our fellow professionals coined the term 'user,' just as doctors refer to those to whom they minister as 'patients.' However, if they ignore the feelings of their supplicants, both doctors and computer designers may be failing to take seriously enough their respective workplace oaths." When we interact with computers, we'd like to be at ease and not at odds; for this interaction to be zestful, our feelings must come into play.

We've seen twenty-six items in the emerging new computing dictionary, twenty-six bright colors that are growing ever brighter. What's more, the computing landscape is not confined to the monochromatic, exhibiting but this or that color. We can create multicolored fabrics, systems that evolve, develop, learn, behave fuzzily, and whatnot. The colors of the computing rainbow join to form the new mosaic of computing.

From adaptive to zestful it seems that computing has never been as exciting and thriving as it is today—as well as complex, and baffling at times, thus echoing Charles Dickens's words in *A Tale of Two Cities*: "It was the best of times, it was the worst of times."

"I never think of the future. It comes soon enough," quipped Albert Einstein during an interview given in 1930. But humans that we are, we are unable not to reflect upon the path ahead: What lies just beyond the horizon? Well, perhaps we will see computing machines that are

Artistic,
  Born,
    Creative,
      Dreamy,
        Emotional,
      Feeling,
  Genteel,
  Humane,
Intelligent,
  Judgmental,
    Keen,
      Lifelike (nay, Living?),
        Moral,
          Noisy (and Nosy),
        Observant,
    Playful,
  Quixotic,
    Resourceful,
      Social,
        Thinking,
          Understanding,
        Vagarious,
    Witty,
  Xenophilous,
  Yakking,
Zealous.

Sound familiar?

Epilogue

Once upon a time, in the faraway kingdom of Adama, there lived a young girl named Adi. Cheerful and ever curious, Adi was the pride and joy of her parents. Her father Val was an accomplished sculptor, known across the kingdom for his magnificent android creations. For her fourth birthday Val had fashioned Adi two wonderful playmates, whom she christened Asi and Mov.

The three of them had become best friends in no time and would pass many a cheerful hour together. So much so, in fact, that Mor—her mother—began to worry. "She spends too much time with Asi and Mov," Mor would complain to Val, "and too little time in the company of humans." To which Val would reply calmly: "She's having a good time, and I see no harm in it; anyway, she'll soon be going to school, where she'll meet other children." And so indeed it came to pass.

Like most children Adi hated school at first, but she soon changed her mind. Maybe this was because she made lots of new friends. Or maybe it was because—like the other children—she was allowed to bring her robotic friends Asi and Mov to school. Or just maybe it was because one of her teachers was her mother. Whatever the reason, Adi was happy.

Adi liked Mor's classes most of all. Her mother taught a class on creativity, where all the children got to do what children are so good at: imagine. The part Adi liked best about these classes was that the students got to use the big computer, called Bim. All the children liked Bim, since he was so eager to please them, and to grant their every wish; besides, he talked funny. (Somehow Adi came to think of Bim as "he," though Asi and Mov just thought of Bim as "it.")

There was that one time when Adi asked Bim to make lots of little robots, like Asi and Mov, and then she created a world for them to live in. Well, to be honest, Bim helped her a little in getting things right.... For two whole weeks she'd come rushing in every morning to see how her world was doing, delighting in all the new lifies—as she called them—that would pop up. Two of the lifies she liked so much, she decided to keep them. One was a weird-looking, birdlike animal, whom she named Dar, and the other, whom she named Win, was a six-legged creature that looked like a small fur ball. Mor told her she'd evolved very lovely creatures, though Adi wasn't quite sure what exactly "evolved" meant; "probably just one of those grown-up words," she thought to herself.

One time Bim had been acting very strangely. He would not comply with Adi's wishes, and he did not talk funny. All the children were very upset, but Mor explained to them that this happens from time to time. "Just like us humans," she said, "computers can also get sick. But don't worry; Bim's very strong, and he'll get better soon." And so it came to pass, just as Mor had said: Bim recovered and was back to his cheerful old self in no time.

Adi didn't like math so much, but luckily Asi and Mov did. Both of them enjoyed doing their math homework, which was really great since they could help Adi with hers.

History, though, was one of Adi's favorites. When the history teacher announced that they'd be going on a field trip to the Computer History Museum in the kingdom's capital, she became so excited that Val and Mor had a hard time convincing her to go to bed the night before.

The visit was a revelation to Adi. Most of the computers still worked, or at least that's what the teacher said. But when she tried talking to them, they wouldn't answer. And none of them could create things the way Bim did. And the old robots, who she was told were the ancestors of Asi and Mov, walked so funny. She tried to teach one of them, who looked a bit more friendly, how to play hopscotch, and she was quite upset when she made no progress. "They're nothing like Asi and Mov," explained the teacher, "so you can't expect too much of them." The visit ended in the hall of images, which contained old twodees of ancient computers. "How funny they look," thought Adi, "Nothing at all like Bim."

It was on that day that Adi decided she wanted to be both a robo sculptor and a teacher when she grew up. Just like mommy and daddy.

Of course, Adi grew up to become something entirely different.

But that's another story.

Coda

If I had wings I'd fly away,
Beyond the realms of night and day.
If I had gills I'd plunge alow,
Where beings brightly be aglow.
If I had piercing cybereyes,
I'd see beyond the web of lies.
If I had memory galore,
Lo the depths I would explore.
If I begot that which does not
Exist upon this earthly lot.
If I can spring such reveries,
Might they be hazy memories,
Of a future that awaits
As we open wide the gates?

Machine Nature

As we open wide the gates
To an infinitude of fates,
Every blink of every eye
Skips bifurcations rushing by,
And breezes of a nova terra
Sound the trumpets of an era,
My reeling mind ascends, transcends:
Whither goeth sapiens?

Further Reading

Below you'll find a small selection of references for each chapter. In addition, I maintain a Web page with links to most of the topics discussed in this book:

http://lslwww.epfl.ch/~moshes/caslinks.html
http://www.cs.bgu.ac.il/~sipper/caslinks.html

Introduction

P. Coveney and R. Highfield, *Frontiers of Complexity: The Search for Order in a Chaotic World*, Faber and Faber, London, 1995.

Chapter 1

C. Darwin, *On the Origin of Species by Means of Natural Selection, or the Preservation of Favoured Races in the Struggle for Life*, John Murray, London, 1859. In point of historical fact one should perhaps refer to the Darwin-Wallace (or Wallace-Darwin) theory of evolution. As pointed out by Darwin himself in the introduction to his book: "I have been urged to publish this Abstract. I have more especially been induced to do this, as Mr Wallace, who is now studying the natural history of the Malay archipelago, has arrived at almost exactly the same general conclusions that I have on the origin of species." Indeed, the works of both Darwin and Wallace had been presented to the Linnaean Society on the same day (July 1, 1858).

D. B. Fogel, *Evolutionary Computation: Toward a New Philosophy of Machine Intelligence*, IEEE Press, Piscataway, NJ, 2d ed., 1999. A good overview, covering a number of subdomains of the evolutionary-computation field.

P. Funes and J. Pollack, "Evolutionary body building: Adaptive physical designs for robots," *Artificial Life*, vol. 4, no. 4, pp. 337–357, Fall 1998. The paper on evolving Lego bridges.

Z. Michalewicz, *Genetic Algorithms + Data Structures = Evolution Programs*, Springer-Verlag, Heidelberg, 3d ed., 1996. A comprehensive volume on evolutionary computation and its applications.

K. Sims, "Artificial evolution for computer graphics," *Computer Graphics*, vol. 25, no. 4, pp. 319–328, July 1991.

M. Sipper, "Notes on the origin of evolutionary computation," *Complexity*, vol. 4, no. 5, pp. 15–21, May/June 1999. A paper that recounts some of the relationships between Darwin's *Origin* and modern evolutionary computation.

M. D. Vose, *The Simple Genetic Algorithm: Foundations and Theory*, The MIT Press, Cambridge, MA, 1999. Concentrates on a popular subdomain of evolutionary computation known as genetic algorithms.

The field of evolutionary computation is highly active these days, not only in academic circles but also among industrials firms. Several annual scientific conferences are held each year, and there are also two research journals published regularly: *IEEE Transactions on Evolutionary Computation* (published by the IEEE) and *Evolutionary Computation* (published by the MIT Press).

Chapter 2

W. Banzhaf, P. Nordin, R. E. Keller, and F. D. Francone, *Genetic Programming—An Introduction: On the Automatic Evolution of Computer Programs and Its Applications*, Morgan Kaufmann

Publishers, San Francisco, CA, 1997. A good introduction to the field.

J. R. Koza, *Genetic Programming: On the Programming of Computers by Means of Natural Selection*, The MIT Press, Cambridge, MA, 1992. The book that launched the field of genetic programming. Thorough and well written, with several sample applications.

J. R. Koza, *Genetic Programming II: Automatic Discovery of Reusable Programs*, The MIT Press, Cambridge, MA, 1994. The second book introduced the idea of automatically decomposing a problem into subproblems, an important step in the solution of complex problems.

J. R. Koza, F. H. Bennett III, D. Andre, and M. A. Keane, *Genetic Programming III: Darwinian Invention and Problem Solving*, Morgan Kaufmann, San Francisco, CA, 1999. The third book is devoted mainly to the evolution of electrical circuits.

As of the year 2000, the field boasts its own research journal: *Genetic Programming and Evolvable Machines* (published by Kluwer Academic Publishers).

Chapter 3

S. Haykin, *Neural Networks: A Comprehensive Foundation*, Macmillan College, New York, NY, 1994. A comprehensive treatment of neural networks from an engineering perspective.

J. Hertz, A. Krogh, and R. G. Palmer, *Introduction to the Theory of Neural Computation*, Addison-Wesley Publishing Company, Redwood City, CA, 1991. For those interested in theoretical aspects of neural computation.

J. J. Hopfield, "Neural networks and physical systems with emergent collective computational abilities," *Proceedings of the National Academy of Sciences, USA*, vol. 79, pp. 2554–2558, April 1982. Harbinger of the revival of artificial neural networks.

W. S. McCulloch and W. Pitts, "A logical calculus of the ideas immanent in nervous activity," *Bulletin of Mathematical Biophysics*, vol. 5, pp. 115–133, 1943.

M. Minsky and S. Papert, *Perceptrons: An Introduction to Computational Geometry*, The MIT Press, Cambridge, MA, 1969.

The book that is considered to have caused much distress to artificial neural-network researchers in the 1970s.

R. Rojas, *Neural Networks: A Systematic Introduction*, Springer-Verlag, Berlin, 1996. A comprehensive introduction to the field of artificial neural networks.

F. Rosenblatt, "The Perceptron: A probabilistic model for information storage and organization in the brain," *Psychological Review*, vol. 65, pp. 386–408, 1958.

D. E. Rumelhart, J. L. McClelland, and the PDP Research Group, eds., *Parallel Distributed Processing: Explorations in the Microstructure of Cognition*, The MIT Press, Cambridge, MA, 1986. A seminal, two-volume work, which showed, among other things, how to overcome the limitations of perceptrons presented by Minsky and Papert.

The field of artificial neural networks is highly active these days, both in academia and in industry. Many annual scientific conferences are held each year, and there are also several research journals published regularly, including: *Neural Computation* (published by The MIT Press), *IEEE Transactions on Neural Networks* (published by the IEEE), and *Neural Networks* (published by Pergamon Press).

Chapter 4

R. A. Brooks, "New approaches to robotics," *Science*, vol. 253, no. 5025, pp. 1227–1232, September 1991. An early paper on the new approach to robotics by a leading researcher in the field.

D. Floreano and F. Mondada, "Evolution of homing navigation in a real mobile robot," *IEEE Transactions on Systems, Man, and Cybernetics—Part B*, vol. 26, no. 3, pp. 396–407, 1996. A paper describing the evolution of obstacle-avoidance behavior in real (Khepera) robots.

S. Nolfi and D. Floreano, *Evolutionary Robotics: The Biology, Intelligence, and Technology of Self-Organizing Machines*, The MIT Press, Cambridge, MA, 2000. A recent book describing the field of evolutionary robotics.

C. Teuscher, E. Sanchez, and M. Sipper, "Romero's odyssey to Santa Fe: From simulation to real life." In M. Jamshidi,

A. A. Maciejewski, R. Lumia, and S. Nahavandi, eds., *Robotic and Manufacturing Systems: Recent Results in Research, Development and Applications*, TSI Press, Albuquerque, NM, 2000, vol. 10 of *TSI Press Series: Proceedings of the World Automation Congress*, pp. 262–267. Also: http://www.teuscher.ch/christof/romero_main.html

R. A. Watson, S. G. Ficici, and J. B. Pollack, "Embodied evolution: Embodying an evolutionary algorithm in a population of robots." In P. Angeline, Z. Michalewicz, M. Schoenauer, X. Yao, A. Zalzala, eds., *1999 Congress on Evolutionary Computation*, IEEE Press, Piscataway, NJ, pp. 335–342, 1999.

B. Webb, "A cricket robot," *Scientific American*, vol. 275, no. 6, pp. 62–67, December 1996. The paper about the cricket-robot experiments.

The journal *Adaptive Behavior* (published by the International Society for Adaptive Behavior) is a good source of information on research in this field.

Chapter 5

B. Kosko, *Fuzzy Thinking: The New Science of Fuzzy Logic*, Hyperion, New York, NY, 1993. A good nontechnical introduction to the field by one of its most prominent researchers.

D. McNeill and P. Freiberger, *Fuzzy Logic*, Simon & Schuster, New York, NY, 1993. Another nontechnical introduction, which also discusses the impact of this technology on American business.

C.-A. Peña-Reyes and M. Sipper, "A fuzzy-genetic approach to breast cancer diagnosis," *Artificial Intelligence in Medicine*, vol. 17, no. 2, pp. 131–155, October 1999.

C.-A. Peña-Reyes and M. Sipper, "Fuzzy CoCo: A cooperative-coevolutionary approach to fuzzy modeling," *IEEE Transactions on Fuzzy Systems*, vol. 9, no. 5, pp. 727–737, October 2001.

L. A. Zadeh, "Fuzzy sets," *Information and Control*, vol. 8, pp. 338–353, 1965. Launched the field of fuzzy logic.

The field of fuzzy logic sees many conferences held each year, as well as a number of research journals, including: *IEEE Transactions on Fuzzy Systems* (published by the IEEE), *Fuzzy Sets and Systems* (published by North-Holland), and *International Journal of Approximate Reasoning* (published by North-Holland).

Chapter 6

E. Sanchez and M. Tomassini, *Towards Evolvable Hardware, Lecture Notes in Computer Science,* vol. 1062, Springer-Verlag, Heidelberg, 1996.

M. Sipper and D. Mange, eds., *IEEE Transactions on Evolutionary Computation: Special Issue—From Biology to Hardware and Back,* vol. 3, no. 3, September 1999. A special issue on evolvable hardware.

M. Sipper, D. Mange, and E. Sanchez, "Quo vadis evolvable hardware?" *Communications of the ACM,* vol. 42, no. 4, pp. 50–56, April 1999. A paper on evolvable hardware.

M. Sipper, E. Sanchez, D. Mange, M. Tomassini, A. Pérez-Uribe, and A. Stauffer, "A phylogenetic, ontogenetic, and epigenetic view of bio-inspired hardware systems," *IEEE Transactions on Evolutionary Computation,* vol. 1, no. 1, pp. 83–97, April 1997. A paper that describes the use of configurable processors for building bio-inspired hardware.

J. Villasenor and W. H. Mangione-Smith, "Configurable computing," *Scientific American,* vol. 276, no. 6, pp. 54–59, June 1997. A good introductory article, with sample applications in the areas of image processing and data encryption.

Chapter 7

F. Jacob, *La logique du vivant: Une histoire de l'hérédité,* Gallimard, Paris, 1970. Original French version of François Jacob's book.

F. Jacob, *The Logic of Life: A History of Heredity,* Princeton University Press, Princeton, NJ, 1993. English translation of Jacob's book.

D. Mange and M. Tomassini, eds., *Bio-Inspired Computing Machines: Toward Novel Computational Architectures,* Presses Polytechniques et Universitaires Romandes, Lausanne, Switzerland, 1998. This book includes a comprehensive description of the Embryonic Electronics project.

M. Ridley, *Genome: The Autobiography of a Species in 23 Chapters*, Fourth Estate, London, 1999. The quotation by Ridley cited in the chapter is drawn from this book (p. 184).

G. Tempesti, D. Mange, and A. Stauffer, "Self-replicating and self-repairing multicellular automata," *Artificial Life*, vol. 4, no. 3, pp. 259–282, 1998. An overview of the Embryonic Electronics project, which uses little electronics jargon.

L. Wolpert, *The Triumph of the Embryo*, Oxford University Press, New York, NY, 1991.

L. Wolpert, R. Beddington, J. Brockes, T. Jessell, P. Lawrence, and E. Meyerowitz, *Principles of Development*, Oxford University Press, Oxford, 1998. A recent and thorough book on embryonic development in nature.

Chapter 8

S. Forrest, S. A. Hofmeyr, and A. Somayaji, "Computer immunology," *Communications of the ACM*, vol. 40, no. 10, pp. 88–96, October 1997. A paper that describes the work carried out by the University of New Mexico group.

J. O. Kephart, G. B. Sorkin, D. M. Chess, and S. R. White, "Fighting computer viruses," *Scientific American*, vol. 277, no. 5, pp. 56–61, November 1997. A paper that describes the work carried out by the IBM team.

Chapter 9

L. M. Adleman, "Molecular computation of solutions to combinatorial problems," *Science*, vol. 266, pp. 1021–1024, November 1994. The paper that launched the field of DNA computing, in which Leonard Adleman solved the traveling salesperson problem (or rather the simplified version with equal distances between cities) for a given seven-city map.

L. M. Adleman, "Computing with DNA," *Scientific American*, vol. 279, no. 2, pp. 34–41, August 1998. In this article Adleman recounts how he came upon the idea of computing with DNA and goes on to describe the details of the actual experiment performed in the lab. The quotation from Adleman cited in the chapter is drawn from this paper.

M. H. Garzon and R. J. Deaton, "Biomolecular computing and programming," *IEEE Transactions on Evolutionary Computation*, vol. 3, no. 3, pp. 236–250, September 1999. A review paper on DNA and molecular computing. Includes a description of molecular evolutionary computation.

D. Harel, *Algorithmics: The Spirit of Computing*, Addison-Wesley Publishing Company, Boston, 2d ed., 1992. A superb introduction to the fundamental questions studied in computing science, including NP-complete problems, such as that of the traveling salesperson.

R. J. Lipton, "DNA solution of hard computational problems," *Science*, vol. 268, pp. 542–545, April 1995. Richard Lipton addressed another one of those tough problems (known as "satisfiability" among computing scientists). He also discussed some of the extant problems that need to be solved in order for the technique to become viable.

Chapter 10

P. W. Anderson, "More is different," *Science*, vol. 177, no. 4047, pp. 393–396, August 1972. Philip W. Anderson's paper, cited at the beginning of the chapter.

M. S. Capcarrère, M. Sipper, and M. Tomassini, "Two-state, $r = 1$ cellular automaton that classifies density," *Physical Review Letters*, vol. 77, no. 24, pp. 4969–4971, December 1996. This paper describes a cellular solution to the majority problem.

L. O. Chua, *CNN: A Paradigm for Complexity*, vol. 31 of *World Scientific Series on Nonlinear Science*, World Scientific, Singapore, 1998. A book on cellular neural networks (CNN).

M. Dorigo, G. Di Caro, and L. M. Gambardella, "Ant algorithms for discrete optimization," *Artificial Life*, vol. 5, no. 2, pp. 137–172, Spring 1999. An overview of work on ant-inspired computing.

W. D. Hillis, *The Connection Machine*, The MIT Press, Cambridge, MA, 1985. The quotation by Hillis cited in the chapter is drawn from this book (p. 139).

M. Sipper, *Evolution of Parallel Cellular Machines: The Cellular Programming Approach*, Springer-Verlag, Heidelberg, 1997.

This book shows how artificial evolution can be applied to evolving cellular computers.

M. Sipper, "The emergence of cellular computing," *IEEE Computer*, vol. 32, no. 7, pp. 18–26, July 1999. A paper that gives an in-depth overview of the field of cellular computing.

G. Theraulaz and E. Bonabeau, "A brief history of stigmergy," *Artificial Life*, vol. 5, no. 2, pp. 97–116, Spring 1999. Includes a description of the simulation of artificial, nest-building wasps.

The quotation by Lipton cited in the chapter is drawn from his paper "DNA solution of hard computational problems" (see references of Chapter 9).

Chapter 11

R. A. Brooks, "Elephants don't play chess," *Robotics and Autonomous Systems*, vol. 6, no. 1–2, pp. 3–15, June 1990.

R. Dawkins, *The Blind Watchmaker*, W. W. Norton & Company, New York, NY, 1986.

J. D. Farmer and A. d'A. Belin, "Artificial life: The coming evolution." In C. G. Langton, C. Taylor, J. D. Farmer, and S. Rasmussen, eds., *Artificial Life II*, Addison-Wesley, Redwood City, CA, 1992, vol. 10 of *SFI Studies in the Sciences of Complexity*, pp. 815–840. Contains a short discussion on the merits of using Lamarckian evolution in artificial systems (see Dawkins's book for a general discussion of Lamarckism).

IEEE Spectrum, vol. 33, no. 5, May 1996. A special issue on artificial eyes, exemplifying the imitation of Nature's widgets by engineers.

Chapter 12

D. C. Dennett, *Darwin's Dangerous Idea: Evolution and the Meanings of Life*, Simon & Schuster, New York, NY, 1995. A modern examination of Darwin's theory of evolution.

G. Egan, *Permutation City*, Harper, New York, NY, 1994.

M. Sipper, "If the milieu is reasonable: Lessons from nature on creating life," *Journal of Transfigural Mathematics*, vol. 3, no. 1, pp. 7–22, 1997. A paper that discusses various issues related to artificial life (includes passages from *Permutation City* that serve to demonstrate the points made).

Chapter 13

I. Harvey, "Creatures from another world," *Nature*, vol. 400, pp. 618–619, August 12, 1999.

B. A. Huberman, P. L. T. Pirolli, J. E. Pitkow, and R. M. Lukose, "Strong regularities in World Wide Web surfing," *Science*, vol. 280, pp. 95–97, April 1998.

T. S. Kuhn, *The Structure of Scientific Revolutions*, University of Chicago Press, Chicago, IL, 1962.

S. Lawrence and C. L. Giles, "Searching the World Wide Web," *Science*, vol. 280, pp. 98–100, April 1998.

R. E. Lenski, C. Ofria, T. C. Collier, and C. Adami, "Genome complexity, robustness and genetic interactions in digital organisms," *Nature*, vol. 400, no. 6745, pp. 661–664, August 1999.

T. S. Ray, "An approach to the synthesis of life." In C. G. Langton, C. Taylor, J. D. Farmer, and S. Rasmussen, eds., *Artificial Life II*, Addison-Wesley, Redwood City, CA, 1992, vol. 10 of *SFI Studies in the Sciences of Complexity*, pp. 371–408. A comprehensive paper on Tierra. The best source of information on Tierra is the Tierran Web site, maintained by Ray: http://www.isd.atr.co.jp/~ray/tierra/index.html.

L. Wolpert, *The Unnatural Nature of Science*, Faber and Faber, London, 1992.

Chapter 14

C. G. Langton, ed., *Artificial Life: Proceedings of an Interdisciplinary Workshop on the Synthesis and Simulation of Living Systems*, vol. 6 of *SFI Studies in the Sciences of Complexity*, Addison-Wesley, Redwood City, CA, 1989. Proceedings of the first artificial-life conference, held in Los Alamos, New Mexico, in 1987.

C. G. Langton, C. Taylor, J. D. Farmer, and S. Rasmussen, eds., *Artificial Life II: Proceedings of the Workshop on Artificial Life*, vol. 10 of *SFI Studies in the Sciences of Complexity*, Addison-Wesley, Redwood City, CA, 1992. Proceedings of the second artificial-life conference, held in Santa Fe, NM in 1990. The quotations from Langton cited in the chapter are drawn from the preface to this volume.

S. Levy, *Artificial Life: The Quest for a New Creation*, Random House, New York, NY, 1992. Though slightly outdated, it is probably the best starting point if you wish to learn more about artificial life.

M. Sipper, "An introduction to artificial life," *Explorations in Artificial Life* (special issue of *AI Expert*), pp. 4–8, Miller Freeman, San Francisco, CA, September 1995. A short introduction to the field. (You may also consult the paper "If the milieu is reasonable: Lessons from nature on creating life," cited in the references of Chapter 12.)

M. Sipper and J. A. Reggia, "Go forth and replicate," *Scientific American*, vol. 284, no. 8, August 2001.

M. Sipper, G. Tempesti, D. Mange, and E. Sanchez, eds., *Artificial Life Journal: Special Issue on Self-Replication*, vol. 4, no. 3, Summer 1998.

J. von Neumann, *Theory of Self-Reproducing Automata*, University of Illinois Press, Champaign, IL, 1966, edited and completed by A. W. Burks.

J. Wyndham "Female of the species." In P. Haining, ed., *The Flying Sorcerers: More Comic Tales of Fantasy*, Orbit Publications, UK, 1997.

There are two research journals devoted exclusively to artificial life: *Artificial Life* (published by The MIT Press) and *Artificial Life and Robotics* (published by Springer-Verlag, Tokyo). Two major conferences are held alternately: Artificial Life (in even-numbered years), ECAL-European Conference on Artificial Life (in odd-numbered years).

Chapter 15

P. J. Denning and R. M. Metcalfe, eds., *Beyond Calculation: The Next Fifty Years of Computing*, Springer-Verlag, New York, NY, 1997. A compendium containing twenty essays written by leading computing scientists. The quotations from Peter J.

Denning, Richard W. Hamming, Mark Weiser and John Seely Brown, and Abbe Mowshowitz are taken from their chapters appearing in this volume.

M. J. Flynn, "Parallel processors were the future ... and may yet be," *IEEE Computer*, vol. 29, no. 12, pp. 151–152, December 1996.

S. Mann, "Wearable computing: A first step toward personal imaging," *IEEE Computer*, vol. 30, no. 2, pp. 25–32, February 1997.

E. M. A. Ronald and M. Sipper, "Why must computers make us feel blue, see red, turn white, and black out?" *IEEE Spectrum*, vol. 36, no. 9, pp. 28–31, September 1999.

E. M. A. Ronald and M. Sipper, "The challenge of tamperproof Internet computing," *IEEE Computer*, vol. 33, no. 9, pp. 98–99, September 2000.

Index

Index

Index

Index

Index

Index

Index

Index

ABOUT THE AUTHOR

Moshe Sipper, Ph.D., is an Associate Professor in the Department of Computer Science at Ben-Gurion University in Israel and a Visiting Professor in the Logic Systems Laboratory at the Swiss Federal Institute of Technology in Lausanne. Dr. Sipper has published close to 100 scientific papers in the field of bio-inspired computing, and is the author of *Evolution of Parallel Cellular Machines: The Cellular Programming Approach*. His Web site is *www.moshesipper. com*.